Advances in Intelligent Systems and Computing

Volume 307

Series editor

Janusz Kacprzyk, Polish Academy of Sciences, Warsaw, Poland
e-mail: kacprzyk@ibspan.waw.pl

About this Series

The series "Advances in Intelligent Systems and Computing" contains publications on theory, applications, and design methods of Intelligent Systems and Intelligent Computing. Virtually all disciplines such as engineering, natural sciences, computer and information science, ICT, economics, business, e-commerce, environment, healthcare, life science are covered. The list of topics spans all the areas of modern intelligent systems and computing.

The publications within "Advances in Intelligent Systems and Computing" are primarily textbooks and proceedings of important conferences, symposia and congresses. They cover significant recent developments in the field, both of a foundational and applicable character. An important characteristic feature of the series is the short publication time and world-wide distribution. This permits a rapid and broad dissemination of research results.

More information about this series at http://www.springer.com/series/11156

Wojciech Zamojski · Jarosław Sugier
Editors

Dependability Problems of Complex Information Systems

Editors
Wojciech Zamojski
Institute of Computer Engineering, Control and Robotics
Wrocław University of Technology
Wrocław
Poland

Jarosław Sugier
Institute of Computer Engineering, Control and Robotics
Wrocław University of Technology
Wrocław
Poland

ISSN 2194-5357 ISSN 2194-5365 (electronic)
ISBN 978-3-319-08963-8 ISBN 978-3-319-08964-5 (eBook)
DOI 10.1007/978-3-319-08964-5

Library of Congress Control Number: 2014943647

Springer Cham Heidelberg New York Dordrecht London

Printed on acid-free paper

Springer is part of Springer Science+Business Media (www.springer.com)

Preface

We are pleased and honoured to present the monograph on *Dependability problems of complex information systems* that includes some original approaches to the selected problems of complex systems dependability.

Contemporary technical systems are integrated compositions of technical, information, organization, software and human (users, administrators and management) resources. Their complexity stems from the applied technical and organizational structures (comprising both hardware and software resources), but even more, from the complexity of the information processes (processing, monitoring, management, etc.) realized in their operational environment. With system resources being dynamically allocated to the on-going tasks, the flow of system events comprising incoming and/or on-going tasks, management decisions, system faults, defensive system reactions, etc. is modelled as a deterministic or/and probabilistic event stream.

Complexity and multiplicity of processes, their concurrency and their reliance on the in-system intelligence (human and artificial) significantly impedes the construction of strict mathematical models and limits the evaluation of adequate system measures. In many cases, analysis of modern complex systems is confined to quantitative studies (e.g. Monte Carlo simulations) which prevent development of appropriate methods of system design and selection of policies for system exploitation. Security and confidentiality of information processing introduce further complications into the system models and the evaluation methods.

The three basic concepts characterizing the newest approach to modelling and evaluation of dependability properties of contemporary systems are discussed in the monograph:

- modelling of the system and its components,
- tasks (functionalities) performed by the system,
- dependability of the system, which is understood as the correct realization of the tasks in the system and in its environment.

Systems - Complex Systems - Computer Systems

Components of the considered class of systems include devices (hardware), procedures for task realization (software), procedures for the system management (operating/management system), and people (users, administrators, operators, service technicians). The system performs tasks set by the users. Each task is defined as the performance of some work or service on time under the prevailing operating conditions. The necessary system resources are allocated to tasks. The process of system resources allocation is dynamic and depends on various system events, such as start of a new task, end of a running one, device failure, program error, decision of the management system, human fault, etc. The system operates in an environment that also is a source of events, such as hostile attacks on the system.

In these terms we define the model of a complex system that can be used to describe a number of modern entities: computer systems, logistics of a discrete transport system, or even such complex systems as a network of web services.

Tasks - Functionalities

A complex system described above, or more precisely its mathematical model, is built to meet specific user-generated tasks. User requests determine the tasks to be realized. In turn, these tasks are realized by invoking a sequence of functionalities necessary to achieve the desired effect. To perform these functions, the system allocates appropriate resources.

Involvement of system resources and system functionalities in the task is time-varying and depends on various system events.

Reliability - Performability - Dependability

The system is reliable if the user tasks are carried out according to the requirements. Complexity of the system structures enables the correct execution of tasks with different efficiencies. Damages to equipment (hardware) and faults (due to software or human errors) interfere with the correct execution of tasks. In many cases, the system incorporates some measures (hardware redundancy, functional redundancy, time redundancy, repair teams or reconfiguration capabilities) to improve system efficient operation and to minimize its losses caused by faults.

Reliability theory focuses on the elements represented as operational/inoperational blocks. In this approach, system is described by a series-parallel structure. Some years ago, an extension to this reliability model was introduced by including consideration of the functional and performance properties of the system components. In this way the class of functional-reliability models was defined. Performability measures reflect both the functional and performance properties (perform-), and reliability (ability) of the system. In recent years, the term "dependability" has become popular, becoming a better known replacement of performability.

Dependability tries to deal with all the mentioned above challenges by employing a multi-disciplinary approach to theory, technology and maintenance of systems working in a real (and very often unfriendly) environment. Dependability studies investigate the system as a multifaceted and sophisticated amalgamation of technical, information and also human resources concentrating on efficient realization of services in such an environment.

The monograph consists of 11 chapters, representing different approaches to the modeling, analysis and evaluation of the dependability properties of the complex information systems. We hope that the collected works will be valuable to scientists, researchers, practitioners and students who work on problems of dependability. We would like to express our sincere gratitude to the authors of the selected works for their excellent research approach and results.

We are very grateful to the Wrocław University of Technology for their support and funding, which made this monograph possible. It sums up the long years of research initiated at Wrocław by Professor Wojciech Zamojski, aimed at adapting the reliability approach to complex computer-based systems. A substantial part of the monograph presents the results of research done under his guidance within the project N N516 475940 "Dependability improvement of complex information systems by reconfiguration", supported by the Polish National Science Centre.

The Editors
Wojciech Zamojski
Jarosław Sugier

Contents

Prediction of the Performance of Web Based Systems

Dariusz Caban and Tomasz Walkowiak

Wrocław University of Technology, Wybrzeże Wyspiańskiego 27, 50-320 Wrocław, Poland
{dariusz.caban,tomasz.walkowiak}@pwr.edu.pl

Abstract. Complex Web based information systems are organized as a set of component services, communicating using the client-server paradigm. The performance prediction of such systems is complicated by the fact that the service components are strongly inter-dependent. To overcome this issue, it is proposed to use simulation techniques. Extensions to the available network simulation tools are proposed to support this. The authors present the results of multiple experiments with web-based systems, which were conducted to develop a model of client-server interactions adequately describing the relationship between the server response time and resource utilization. This model was implemented in the simulation tools and its accuracy verified against a testbed system configuration.

Keywords: complex information systems, Web based systems, performance assessment, network simulation.

1 Introduction

Accurate prediction of the performance of a web based system, by means of simulation, is in general quite unlikely: there are too many factors that can affect it. Moreover, a lot of these factors are unpredictable, being specific to some unique software feature. This can be overcome in case of predictions made when the system is already production deployed. In this situation, a lot of system information can be collected on the running system. This information can be used to fine tune the simulation models.

Of course, normally this is not useful – the performance can be directly measured in the running system, with no need to recourse to simulation [8]. Sometimes, it is necessary to change the deployment of a running system, either to overcome changes in the demand for service or to overcome some dependability or security issues [3]. Redeployment of service components onto the available hosts changes the workload of the various servers. In consequence some of them are over-utilized and cannot handle all the incoming requests, or handle them with an unacceptable response delay. It is very difficult to predict these side-effects. One of the feasible approaches is to use simulation techniques: to study what are the possible effects of such a change.

Available network simulators are usually capable of analyzing the impact of reconfiguration on the accessability of the services, the settings of the network devices and on security [5,6]. The simulators can predict transmission delays and traffic congestions – that is natural, since it is their primary field of application. They have a very

W. Zamojski and J. Sugier (eds.), *Dependability Problems of Complex Information Systems*,
Advances in Intelligent Systems and Computing 307, DOI: 10.1007/978-3-319-08964-5_1

limited capability to simulate tasks processing by the host computers. It is proposed to overcome this limitation by implementing an empirically validated model of service responses that takes into account the computing resources needed to process requests, models that predict processing delays dependent on the number of concurrently serviced requests [13,14].

The main part of this presentation is dedicated to determining these models and demonstrating their accuracy. We also present some insight into the metrics that are used to characterize the performance of web based systems.

2 Web Based Systems

We consider a class of information systems that is based on web interactions, both at the system – human user (client) interface and between the various distributed system components. This is fully compliant with the service oriented architecture, though it does not imply the use of protocols associated with SOA systems. On the other hand, the applicability of the model is certainly not limited to the service oriented systems. In fact, it encompasses practically all the system architectures utilizing the request-response interactions.

2.1 Simple Web Server Architecture

The simplest example of a web based system consists of a single service, handling a stream of requests coming from multiple clients via Internet. There are three important aspects to modeling this class of systems: infrastructure hosting the service, handling of service requests, client expectations and behavior. All of these have significant impact on the observed system performance.

Host and Network Resources

The service is deployed on a computing host which is connected to the client machine via a network. This deployment determines specific resources available to the service, both in terms of communication throughput and computing power. This deployment has a very significant impact on the service performance, especially the response time.

There is just one communication parameter of significance – the maximum throughput derived from the link bandwidths and the protocols in use. In most practical situations, that we have analyzed, this factor has a very limited impact on the web based systems. In modern installations, the computing resources usually determine the system performance.

The computing resources that need to be considered include the processor speed, available memory, storage interfacing capabilities. Moving a service from one location to another, the available resources change. In consequence, the service performance is affected. This is usually determined by benchmarking the service. To some extent, it can be observed via monitoring of the production system.

It should be noted that the service performance is affected not only when it is re-deployed on a different host. Similar effect is observed, when multiple applications are deployed on the same host. In this case the computing resources are shared by the services, affecting their performance. When trying to predict the web system characteristics, this factor has also to be accounted for.

Client – Server Interactions

The basis of operation of all the web oriented systems is the interaction between a client and a server. This is in the form of a sequence of requests and responses: the client sends a request for some data to the server and, after some delay, the server responds with the required data. The time that elapses from the moment the client sends the request until it receives the response is called the response time.

The response time depends on a number of different factors. As already discussed, it depends on the service deployment and sharing of resources. Just as significantly, specific requests may require different amount of processing. A typical workload is a mixture of different requests. A common approach to load (traffic) generation techniques is based on determining the proportion of the various tasks in a typical server workload, and then mixing the requests in the same proportion [7, 12]. Thus, even in the simple situation, where the response is generated locally by the server, it has an unpredictable, random factor.

Actually, the server response time is strongly related to the client behaviour, as determined by the request-response interaction. Such factors as connection persistence, session tracking, client concurrency or client patience/think times have a documented impact on the reaction. For example, it has been shown in [10] that if user will not receive answer for the service in less than 10 seconds he or she will probably resign from active interaction with the service and will be distracted by other ones.

Let's consider the model used in these simple interactions in more detail. The simplest approach is adopted by the software used for server/service benchmarking, i.e. to determine the performance of computers used to run some web application. In this case, it is a common practice to bombard the server with a stream of requests, reflecting the statistics of the software usage (the proportion of the different types of requests, periods of burst activity, think times, etc.). Sophisticated examples of these models of client-server interaction are documented in the industry standard benchmarks, such as the retired SPECweb2009 [12].

The important factor in this approach is the lack of any feedback between the rate of requests and the server response times. In other words, the client does not wait for the server response, but proceeds to send further requests even if the response is delayed. Fig. 1 shows the results of experiments performed on a typical server application exposed to this type of traffic. Fig. 1 a) presents the changes in the response time, depending on the rate of requests generation. It should be noted that the system is characterized by three distinct ranges in the requests rate.

Up to approximately 35 requests per second, the response time very slowly increases with the rate of requests. This is the underutilization range, where the server processing is not fully utilized: the processor is mainly idle and handles requests immediately on arrival. There is a gradual increase in the response time due to the increased probability of requests handling overlapping.

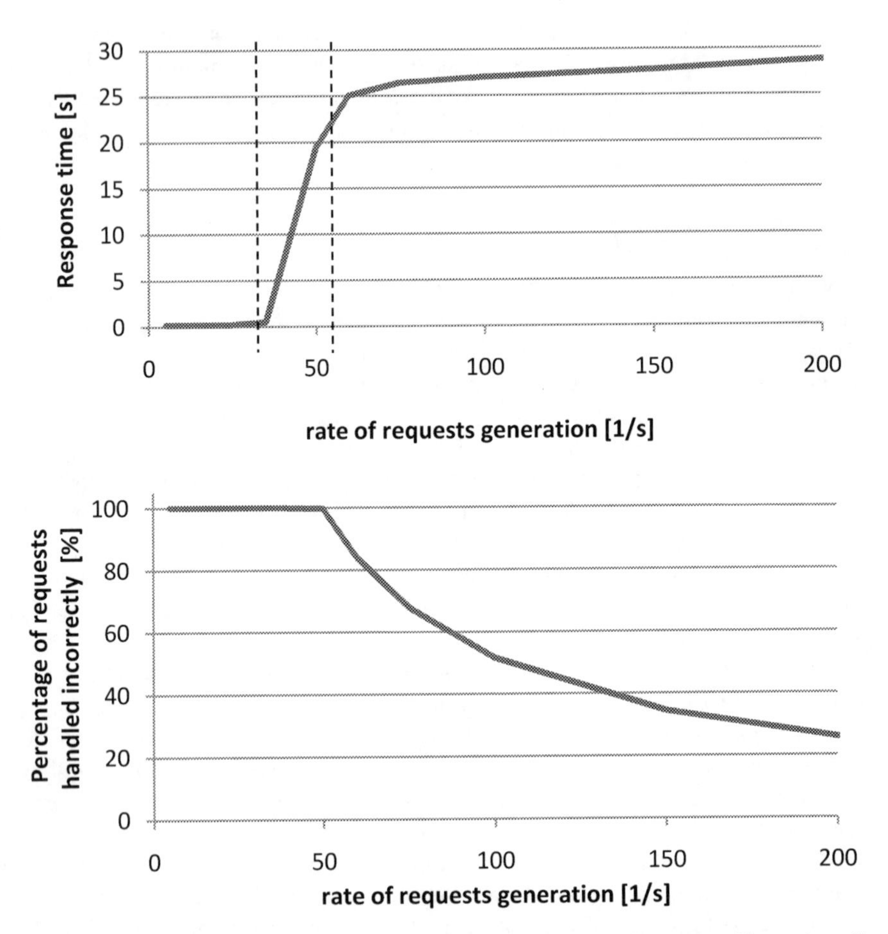

Fig. 1. The performance of an off-the-shelf web service under varying rates of incoming client requests: a) the upper graph shows the response time, b) the lower – the erroneous responses

When the requests rate is higher the processor is fully utilized, the requests are queued and processed concurrently. The increase in the response time is caused by the concurrently handled requests. This range is very narrow, since any significant increase in average requests rate causes the service to be overloaded. Further increase in the request rate does not increase the number of correctly handled ones. Thus, the response time remains almost constant. On the other hand, the percentage of requests handled incorrectly increases proportionately to the request rate. This is illustrated in Fig. 1 b).

Client Models Reflecting Human Reactions

The real behaviour of clients differs significantly from the model discussed so far. In fact, the client sends a burst of related requests to the server, then it waits for the server to respond and, after some "think" time for disseminating the response, sends a new request. Fig. 2 illustrates the timing diagram of such a client.

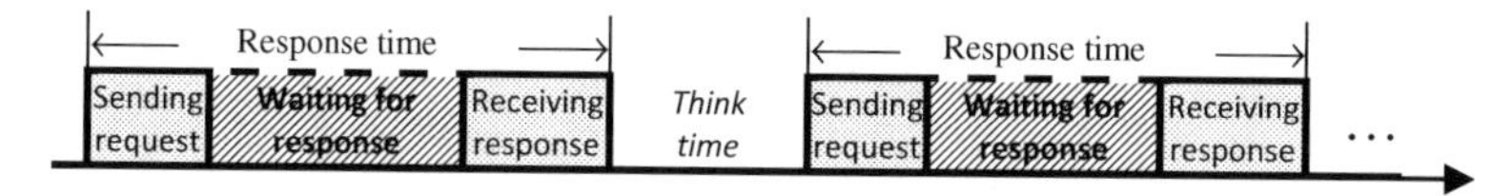

Fig. 2. Client traffic model reflecting request-response sequence and think time

This type of model is implemented in a number of traffic generators available both commercially and in open-source (Apache JQuery, Funkload). The workload is characterized by the number of concurrent clients, sending requests to the server. The actual requests rate depends on the response time and the think time. The model implies that the request rate decreases when the service responds with longer delays (i.e. from the client perspective, the time it waits for the response increases).

This model assumes that the proportion of tasks in a workload does not change significantly due to response delays and error-responding. It does not assume any information on the semantics of client-server interactions. In effect, this produces a mix of tasks, in no way connected to the aims of the clients. The description of client behaviour can be improved if we have a semantic model of client impatience, i.e. how the client reacts to waiting for a server response. Currently, this is modeled very simplistically by setting a threshold delay, after which the client stops waiting for the server response and starts another request. A more sophisticated approach would have to identify the changing client perspective caused by the problems in accessing a service, e.g. a client may reduce the number of queries on products, before deciding to make a business commitment, or on the other hand, he may abandon the commitment. These decisions could significantly influence the workload proportions.

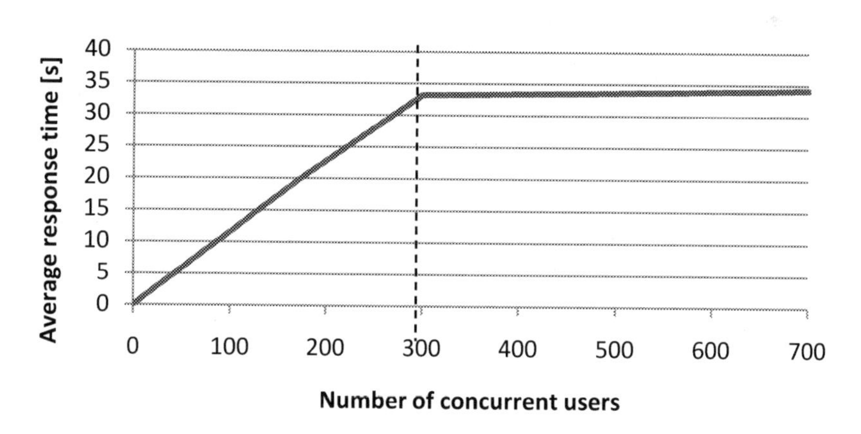

Fig. 3. Average service response when interacting with various number of concurrent clients

Fig. 3 shows how the response time depends on the number of concurrent clients. In this case we have set the "think" time to 0, i.e. a new request is generated by the client directly on receiving the response to a previous one. Quite interestingly, the server operates practically only in the normal utilization range, until it reaches the maximum number of clients that it can handle correctly (roughly 300 clients in Fig. 3).

2.2 Distributed Web Services Architecture

So far, the considered model consisted just of one service, handling all the end-user requests. In a more complex system, the clients interact multiple front-end business services. Furthermore, these services request assistance from other services when computing responses. These interactions determine a network of complementary services (called service components), which communicate with each other using the request-response paradigm.

Service Choreography

The system analysis has to consider the various tasks initiated by the client. In a typical web application, these tasks can exercise the server resources in a wildly varied manner: some will require serving of static web pages, some will require server-side computation, yet others will initiate database transactions or access to remote web applications.

It is assumed that the analyzed web services are described by the choreography description, using one of the formal languages developed for this purpose (we consider WS-CDL and BPEL [11,14] descriptions). This description determines all the sequences of requests and responses performed by the various service components, described in the choreography. Fig. 4 presents a very simple example of service choreography. It should be noted that the choreography determines the sequences of requests and responds at all the interfaces between the service components.

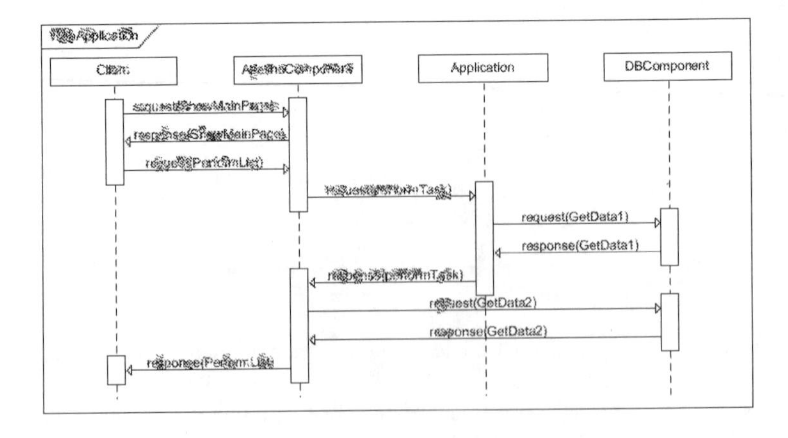

Fig. 4. An example of a simple service choreography

It also places some constraints on the client model. In 2.1, we have assumed that the client sends a random mixture of requests to the web system. Fig. 4 shows that in this specific example the "ShowMainPage" request is followed by the "PerformList" one. Thus, it is not just a random sequence of them. In this approach, the random mixing is performed on the alternative system usage scenarios, instead.

The model of request-response sequences, formulated for client-service communication, is also applicable to interactions between the web service components. In this case one component becomes the client of another. The same timing phenomena can be observed. The client component usually has a built-in response time-out period which corresponds to the end-user impatience time. The significant difference is that, in this case, the choreography description defines the reaction of the client component. Thus, the client impatience model is fully determined, derived from this description.

System Deployment

The service components are deployed on a network of computers. This underlying communication and computing hardware is abstracted as a collection of interconnected computing hosts. System configuration is determined by the deployment of service components onto the hosts. This corresponds to the subsets of services located at each one. The deployment clearly affects the system performance, as it changes the communication and computational requirements imposed on the infrastructure.

The problem of predicting the impact of configuration changes is not trivial. Directly, response times depend on the concurrent load of each host. The greater the number of concurrently handled requests at a host, the slower is the response processing (due to resource sharing). If all resources of a host are already dedicated, a new request has to be queued further increasing the response times. These response delays from one service component propagate to others, affecting both their response times and the workload (numbers of handled requests).

In fact, this is the main application field of the system performance simulation techniques.

3 System Performance Characteristics

There are various approaches to characterizing the quality of the web based systems. Basically, the performance can be assessed in three aspects: their capability to provide responses in the desired timespan, the capability to respond correctly (with possibly few errors), and the ability to handle large, cumulated workloads. We consider the measures directly relating to these service properties.

3.1 Average Service Response Time

The response time is defined as the time that elapses from the moment a client starts sending a request until the response is complete transmitted back to it. This was already discussed in 2.1. The service as a whole is characterized by the response times observed from the user perspective only, i.e. responses to requests sent by the end-user clients.

The average response time is computed over a mixture of user requests, characteristic for the system workload. If the system responds with an error code, the response time is excluded from computing the average. These are not taken into account to prevent false observation of responses speed-up, when the system is overloaded and responding with multiple errors.

The average response time strongly depends on the rate of service requests, as illustrated in Fig. 1 and 3. In case of web services consisting of multiple distributed components, this interdependence is similar in character though different in the observed ranges and scales. To obtain a single value characteristic, a typical request rate has to be used for assessment.

3.2 Service Availability

Availability is normally defined as the probability that a system is operational at a specific time instant [1]. This implies that the system may break down and become inoperational, which is certainly applicable to the web based systems. In these considerations, we assume that the system is operational when we compute its performance characteristics. For this reason, the term "service availability" may be misleading in this case. Instead, we consider availability to be the probability that a request is correctly responded to. It is assessed as the number of properly handled requests n_{ok} expressed as a percentage of all the requests n over a sufficiently long time of operation t :

$$A = \lim_{t \to \infty} \frac{n_{ok}(t)}{n(t)} \tag{1}$$

This yields a common understanding of availability used in the web services community.

The service availability changes with the rate of requests sent to the system. Until the system becomes overloaded the number of error responses should be negligible. It implies that the service availability needs to be assessed for a typical workload, similarly to the response time.

3.3 Maximum System Throughput

The maximum system throughput is defined as the maximum value of incoming requests rate that can properly be handled. This can be determined by:

- assuming specific threshold values of the response time and service availability;
- assessing the two request rates corresponding to these thresholds;
- finding the minimum of the two request rates.

Such approach is not very convenient, since it always requires a clear understanding of the acceptable threshold values. In practical terms, this is always viewed with some uncertainty. A simpler technique, though sacrificing some precision, is to fix the maximum throughput at the value of requests rate midpoint in the range between under- and over-utilization. This value is also very near the point, where service availability begins to decrease rapidly (Fig. 1b).

4 Performance Prediction Using Network Simulation Techniques

There is a large number of network simulators available on the market, both open-source (ns3, Omnet+, SSFNet) and commercial. Most of them are based on the package transport model – simulation of transport algorithms and package queues [5,6]. What they lack is a comprehensive understanding of the computational demands placed on the service hosts, and how it impacts the system performance. For this reason, they cannot be directly used to predict the impact of service components deployment on system performance. The simulators need to be extended, by writing special purpose queuing models for predicting tasks processing time, based on resource consumption [2,13].

Response time prediction in simulators is based on the proper models of the end-user clients, service components, processing hosts (servers), network resources. The client models generate the traffic, which is transmitted by the network models to the various service components. The components react to the requests by doing some processing locally, and by querying other components for the necessary data (this is determined by the system choreography, which parameterizes both the client models and the service component models). The request processing time at the service components is not fixed, though. It depends on the number of other requests being handled concurrently and on the loading of other components deployed on the same hosts.

The simulator needs a number of parameters that have to be set to get realistic results. These parameters are attributed to the various models, mentioned above. In the proposed approach we assume that it is possible to determine the values of these parameters in a running environment. Thus, the technique has limited usefulness, if there is no such data (before the system is initially deployed).

The models should be fairly simple, describing the clients and service components. They should accurately predict changes that may occur when the deployment of service components is modified. Then, simulating the target configuration with these parameters should provide reliable predictions of the web service performance after redeployment.

4.1 Virtual Testbed Environment

A proper model of client-server interactions is the basis for accurate simulation of the system. For this reason, a number of testbed experiments have been conducted to capture the realistic timing characteristics that can be abstracted into a simple model. For this purpose, we have set up a testbed, consisting of a network of virtual machines running the appropriate servers (Apache, IIS, Tomcat, MySQL). The servers run PHP scripts, which can accurately mimic service components. The application is exposed to a stream of requests, generated by a client application (a Python script written by the authors).

The available processor resources are monitored via the virtualization hypervisor to ensure that the traffic generation programs do not compete for the resources with the system software (which would lead to unrealistic results).

4.2 Server Response Prediction

Basic Model

The client-server interaction is paramount to the proper simulation of a complex web service. The analysis of the behaviour of typical servers led to the formulation of a basic model that is used in simulation.

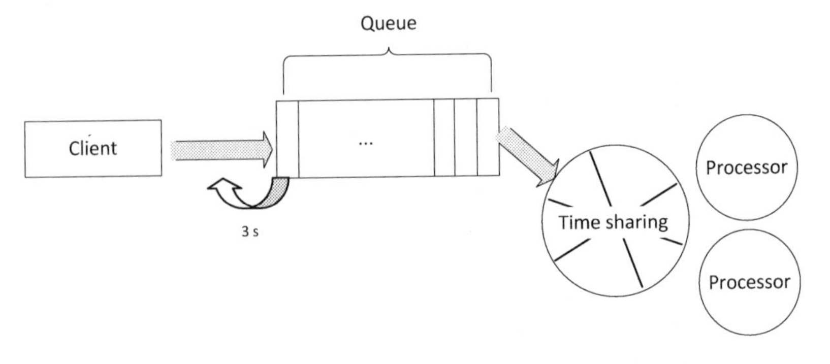

Fig. 5. Basic model of a web service

The basic model, as presented in Fig. 5, consists of four elements: the retransmission buffer, the FIFO style waiting queue, the circular buffer and a set of processors.

The retransmission buffer models the process of establishing TCP connection by a client if a server is not responding. One can observe that connections are established within a discrete time delays: 0, 3, 9, 21, .. seconds. This is implementation of the TCP exponential backoff mechanism, introduced by Jacobson 25 years ago [4] and analyzed in details in many papers, for example in [9].

In the proposed model, the retransmission buffer is working as follows:

1. If the number of processed requests is larger than a given value $N_{\max}$ then the request is rejected within a few ms (a random value).
2. The client waits for a given time period (Δt) for the FIFO (next) queue to accept a request. If it not accepted, then goes to step 3 then it proceeds to step 3.
3. The timeouts parameters are updated:

$$\Delta t = 3 \cdot \Delta t$$
$$t_d = 2 \cdot t_d \;\; + 3s \tag{2}$$

4. The client is paused for t_d seconds.
5. If the time elapsed from the begging of request proceeding is longer than a client timeout ($t_{timeout}$) the request is rejected; if not the client repeats the procedure from step 2.

   The initial values of timeouts are as follows: Δt =0.0125s, t_d =0s.

The waiting queue models requests waiting for execution by the server. It works according to FIFO regime and has only one parameter: its length (N_{FIFO}).

Handling of requests is done by executing a given task or tasks, depending on the requests. It is done in time sharing manner and modelled by the circular buffer. In reality concurrent execution is achieved by switching the processors between different tasks. In general it works as follows:

1. If the circular buffer is not full the request is removed from the end of the waiting queue and moved to the circular buffer and execution of a task defined by a request starts.
2. Each task from the circular buffer has access to a processor (from the set of available one) for a time slice.
3. The task is finished (and removed from the time sharing buffer) when the sum of time slices is larger than the execution time required to process the given request.

In case when just one task is being executed on a given host, the task execution time depends on the host performance described by the parameter *performance*(h) and the task complexity (parameter *tc*()):

$$et(task) = \frac{tc(task)}{performance(h)}. \tag{3}$$

In case more than one task being executed concurrently, the algorithm is more complicated. Let $\tau_1, \tau_2, ..., \tau_e$ be the time moments when some tasks are starting or finishing execution on a host h. Let $number(h,\tau)$ denote the number of tasks being processed (active tasks in circular buffer) at time τ on host h, and $ncores$ the number of processor cores. Therefore, the time when a task finishes its execution has to fulfil the following rule:

$$\sum_{k=2}^{e} (\tau_k - \tau_{k-1}) \frac{performance(h)}{number(h)/ncores} = tc(task). \tag{4}$$

Therefore, the overall processing time is equal to:

$$et(task) = \tau_e - \tau_1. \tag{5}$$

The drawback of the above approach is the fact that it generates an excessive number of events when a large number of tasks are handled concurrently. This is due to the fact that every new request changes the estimated time to finish for each request being executed at this moment. Therefore, we have introduced a heuristic algorithm [13] that prevents the generation of a new event if the previous one (for the same host) was close enough (the time difference is smaller than some threshold).

Implementation of this model allows calculating the processing time of each request as a sum of times spent in the first two queues and its execution time (equations (3,4)).

Basic Model Validation

To verify correctness of the basic model of a web service we have compared simulation results with real Apache server behavior. The results for concurrent clients are shown in Fig. 6.

The results are very accurate considering that we are approximating the complex behaviour of a software component with just a few parameters. The parameters characterize: the host performance, the task complexity, the length of time sharing buffer, the length of a wait queue and maximum number of processed requests (seems to be set to 1000 for most of the web servers).

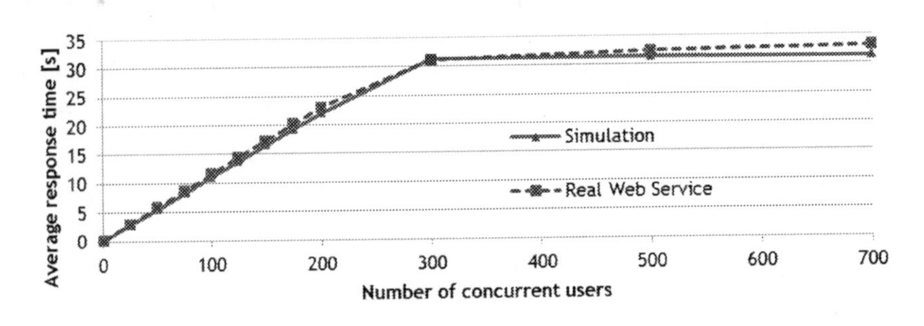

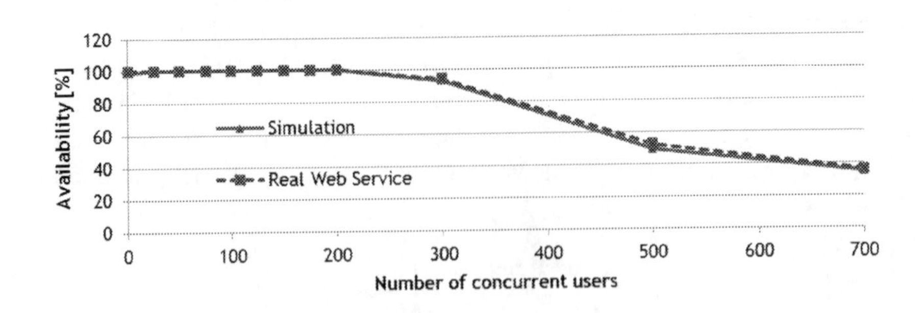

Fig. 6. The performance of a real Apache web server (dashed line) and simulated one (solid line): a) the response time, b) the availability

These parameters could be easily obtained by a simple tests on a real system (the host performance, the task complexity) or from configuration files of the Apache server (*MaxClients* parameters defines the length of the circular buffer) or are predefined by a type of web server (like the length of a wait queue and maximum number of processed requests).

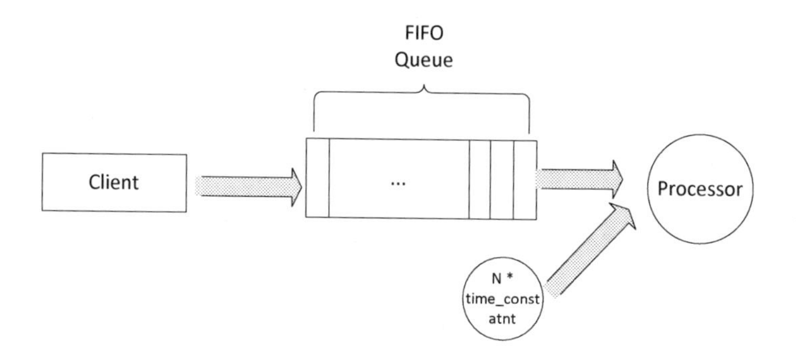

Fig. 7. Simplified service model

Basic Model Modification

In case of some types of servers, particularly some databases and Microsoft IIS, the basic model can be simplified. In these servers, it is not necessary to use the retransmission and circular buffers. The servers can be modelled just by one limited length FIFO queue. In these servers, all requests above the length of the FIFO queue are rejected immediately. Due to simplicity of the model results of simulation for IIS web server are very similar to a real system (in case of response time, it is less than 2%).

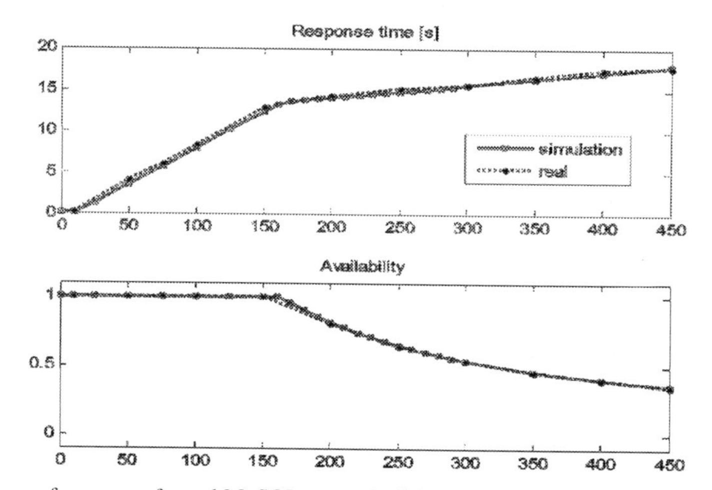

Fig. 8. The performance of a real MySQL server (solid line) and simulated one (dashed line): a) the response time, b) the availability

In case of database systems, as it can be noticed on Fig. 8, there is a constant increase in response time when the server is overutilized. We propose to model it by adding a task which consumes some amount of processor power. The execution time of that additional task is proportional to the number of processed requests:

$$et(request) = N \cdot time\_const \tag{6}$$

The results of simulated and real MySQL server response times are presented in Fig. 8.

4.3 Interaction with Other Services

The operation of all the web based applications is based on the interaction between services. Therefore it is important to model how services process requests that require calls to other service components.

In case of services that follow the basic model (for example Apache, Tomcat), external calls have an influence on the circular buffer. When a task is waiting for an answer from another service (the request thread is in wait state), the place in the circular buffer is used but the processor is not. Therefore, the number of active requests ($number(h,\tau)$) is decreased when a requests starts an external call and increased when the response is received. Such behavior results in a situation that the whole circular buffer is used, so new requests are waiting in FIFO queue whereas the service is not using a processor.

In case of the modified model (without circular buffer) like IIS, the requests waiting for external service response are not using the processor. So, new requests from the FIFO queue can be processed. When the response from the external service arrives, the task is placed in an additional FIFO queue. Therefore, the model for web services without circular buffer uses two FIFO queues (Fig. 9). The processor is processing requests from the two queues alternately.

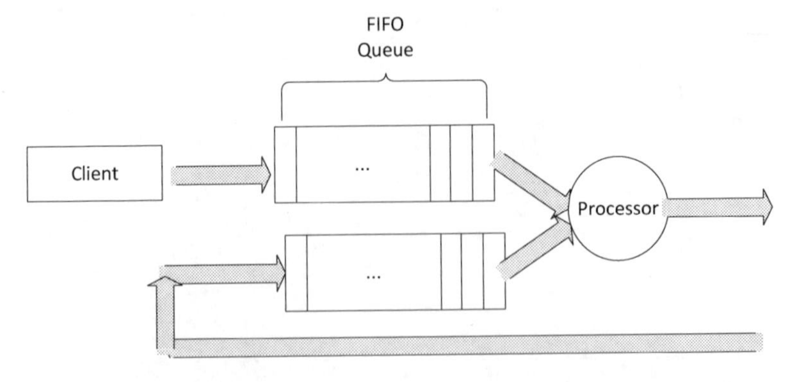

Fig. 9. Simplified model for services interacting with other components

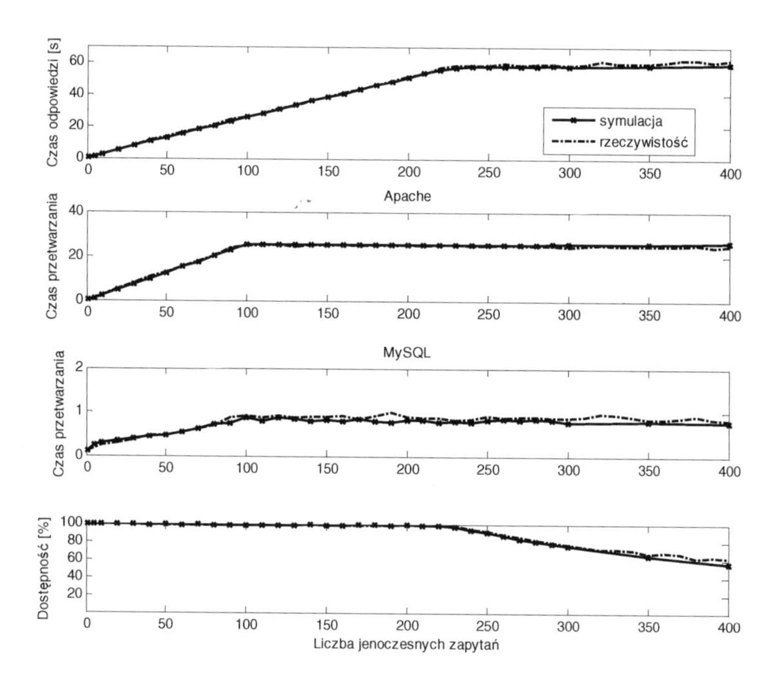

Fig. 10. Results for two layer web system

4.4 Services Deployed on the Same Host

The deployment of multiple services on the same host leads to time-sharing of processor time between them. Each of the active service components deployed on a given hosts gets proportionate access to the processor. To model such situations, we have to add a time sharing queue presented in 4.2 to all hosts regardless the type of used service model.

For the basic model it results in modification of the formula (4) to:

$$\sum_{k=2}^{e}(\tau_k - \tau_{k-1})\frac{performance(h)}{number(s)/ncores \cdot nactiveservers(h)} = tc(task). \tag{7}$$

It results in an increase of the time moments $\tau_1, \tau_2, ..., \tau_e$ since they have to include changes when the number of active services changes.

In case of the simplified model, the time sharing has to be included in the model in a similar way as for the basic one, i.e. the time when a task finishes its execution has to fulfil the following rule:

$$\sum_{k=2}^{e}(\tau_k - \tau_{k-1})\frac{performance(h)}{nactiveservers(h)/ncores} = tc(task). \tag{8}$$

To verify the correctness of the proposed modifications in service models we have performed a set of tests analyzing a simple system with an Apache server and a MySQL database placed on the same host. Results presented in Fig. 10 show that the modified models give results that are very close to the real system behavior.

4.5 Models Based on Service Choreography

The key feature during simulation is to calculate the response times to the end users. The user initiates the communication requesting execution of some tasks on a host. This may require sending a request to another host or hosts. After executing the task the host responds to the requesting service, and finally the user receives the response. Requests and responses of the tasks form a sequence, according to the service choreography. Let's assume that the choreography for some user c_i is given in Fig 4. It can be described in the functional form as:

$$u := t_1\left(t_2\left(t_3(\ )\right), t_4(\ )\right) \tag{9}$$

i.e. execution of user chorography u consists of execution two tasks t_1 and t_4, whereas execution of task t_1 requires calls to task t_2 which calls t_3. Fig. 11 presents the same choreography, with references to the corresponding tasks.

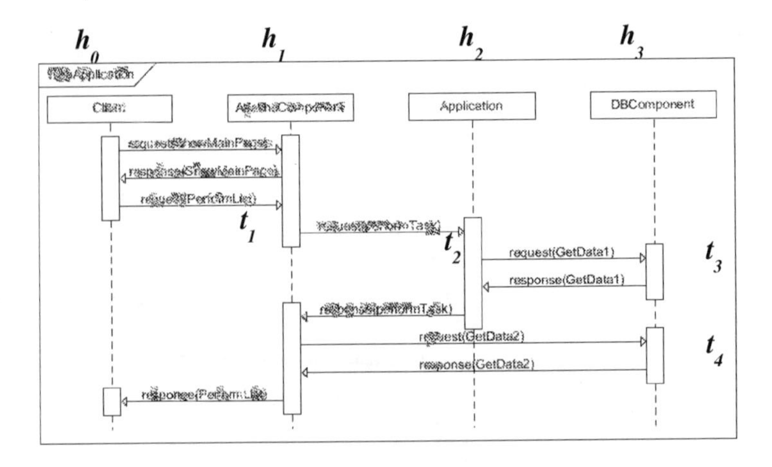

Fig. 11. An example of a service choreography with annotated tasks and hosts deployment

The user request processing time is equal to the time of communication between hosts on which each task is placed and the time of processing of each task. Therefore, for the considered choreography (assuming the deployment of tasks to hosts presented in Fig. 11) the user request processing time is equal to:

$$\begin{aligned} urpt(u) &= com(h_0, h_1) + pt(z_1') + com(h_1, h_2) + pt(z_2') + com(h_2, h_3) + pt(z_3) \\ &+ com(h_3, h_2) + pt(z_2'') + com(h_2, h_1) + pt(z_1'') + com(h_1, h_3) + pt(z_4) + \\ &com(h_3, h_1) + pt(z_1''') + com(h_1, h_0) \end{aligned} \tag{10}$$

where $com(h_i,h_j)$ is the time of transmitting the requests from host h_i to h_j, and $pt(task)$ is the time of processing the task on the given host (i.e. the host that the corresponding service component was deployed to). The processing time consists of the time spent in server queues and the task execution time. It can be calculated by simulation using the presented models.

The communication times in the equation (10) correspond to delays introduced by the network. In almost all modern information systems the local network throughout is high enough, so there is no relation between the number of tasks being processed in the system and the network delay. There are exceptions to this rule, especially in media streaming systems. We propose to model the time of transmitting the requests from host h_i to h_j by independent random values:

$$delay(h_i, h_j) = TNormal(mean, mean \cdot 0.1), \qquad (11)$$

where *TNormal*() denotes the truncated Gaussian distribution (bounded below 0).

5 Conclusions

Performance of the web based information systems is nowadays of utmost importance [8]. Business relies heavily on the high availability of services. Thus, there is a clear need of accurate tools for predicting this performance.

The proposed method of prediction, based on customized network simulation, provides sufficient accuracy. At the same time it does not require very expensive testbed installations that are often used for this purpose. Thus, it is a very promising technique.

The simulation models require a limited number of systems parameters. They make use of knowledge of the service choreography. The approach is particularly well suited when it is necessary to change the deployment of service components in an existing installation. Simulating the expected performance before making the modifications may provide significant guidelines to the choice of optimal reconfiguration.

The technique has limited application to predicting the performance of a system during its development. In this case, the model parameters cannot be observed. Guessing the values of these parameters does not provide sufficiently accurate information to perform meaningful simulation.

Acknowledgement. The presented work was funded by the Polish National Science Centre under grant no. N N516 475940.

References

1. Barlow, R.E.: Engineering Reliability. ASA-SIAM Series on Statistics and Applied Probability (1998)
2. Caban, D., Walkowiak, T.: Service availability model to support reconfiguration. In: Zamojski, W., Mazurkiewicz, J., Sugier, J., Walkowiak, T., Kacprzyk, J. (eds.) Complex Systems and Dependability. AISC, vol. 170, pp. 87–101. Springer, Heidelberg (2012)

3. Caban, D., Walkowiak, T.: Preserving continuity of services exposed to security incidents. In: Proc. The Sixth International Conference on Emerging Security Information, Systems and Technologies, SECURWARE 2012, Rome, August 19-24, pp. 72–78 (2012)
4. Jacobson, V.: Congestion avoidance and control. ACM CCR 18(4), 314–329 (1988)
5. Lavenberg, S.S.: A perspective on queueing models of computer performance. Performance Evaluation 10(1), 53–76 (1989)
6. Liu, J.: Parallel Real-time Immersive Modeling Environment (PRIME), Scalable Simulation Framework (SSF). User's manual. Colorado School of Mines Dept. of Mathematical and Computer Sciences, `http://prime.mines.edu/`
7. Lutteroth, C., Weber, G.: Modeling a Realistic Workload for Performance Testing. In: 12th International IEEE Enterprise Distributed Object Computing Conference (2008)
8. Miller, L.C.: Application Performance Management for Dummies, Riverbed Special edn. John Wiley & Sons, Hoboken (2013)
9. Mondal, A., Kuzmanovic, A.: Removing Exponential Backoff from TCP. ACM SIGCOMM Computer Communication Review 38(5), 19–28 (2008)
10. Nielsen, J.: Usability Engineering. Morgan Kaufmann, San Francisco (1994)
11. Pasley, J.: How BPEL and SOA are changing Web services development. IEEE Internet Computing Magazine 9, 60–67 (2005)
12. SPECweb2009 Release 1.20 Benchmark Design Document version 1.20. SPEC (2010), `http://www.spec.org/web2009/docs/design/SPECweb2009_Design.html`
13. Walkowiak, T.: Information systems performance analysis using task-level simulator. In: Proc. DepCoS – RELCOMEX 2009, pp. 218–225. IEEE Computer Society Press (2009)
14. Walkowiak, T., Michalska, K.: Functional based reliability analysis of web based information systems. In: Zamojski, W., Kacprzyk, J., Mazurkiewicz, J., Sugier, J., Walkowiak, T. (eds.) Dependable Computer Systems. AISC, vol. 97, pp. 257–269. Springer, Heidelberg (2011)

Modelling Uncertain Aspects of System Dependability with Survival Signatures

Frank P.A. Coolen[1] and Tahani Coolen-Maturi[2]

[1] Department of Mathematical Sciences, Durham University, Durham, United Kingdom
[2] Durham University Business School, Durham University, Durham, United Kingdom
{frank.coolen,tahani.maturi}@durham.ac.uk

Abstract. The survival signature was recently introduced to simplify quantification of reliability for systems and networks. It is based on the structure function, which expresses whether or not a system functions given the status of its components. In this paper, we show how a straightforward generalization of the structure function can provide a suitable tool for scenarios of uncertainty and indeterminacy about functioning of a system for the next task. We embed this generalization into the survival signature, leading to a more flexible tool for quantification of the system reliability and related measures of dependability.

1 Introduction

Mathematical theory of reliability has been well established since the middle of the twentieth century, with main focus on the functioning of a system given the functioning, or not, of its components and the structure of the system. The mathematical concept which is central to this theory is the structure function. For a system with m components, let state vector $\underline{x} = (x_1, x_2, \ldots, x_m) \in \{0,1\}^m$, with $x_i = 1$ if the ith component functions and $x_i = 0$ if not. The labelling of the components is arbitrary but must be fixed to define $\underline{x}$. The structure function $\phi : \{0,1\}^m \rightarrow \{0,1\}$, defined for all possible $\underline{x}$, takes the value 1 if the system functions and 0 if the system does not function for state vector $\underline{x}$. Mostly attention is restricted to coherent systems, for which $\phi(\underline{x})$ is not decreasing in any of the components of $\underline{x}$, so system functioning cannot be improved by worse performance of one or more of its components. It is usually assumed that $\phi(\underline{0}) = 0$ and $\phi(\underline{1}) = 1$, so the system fails if all its components fail and it functions if all its components function. When functioning of a system is considered over time, taking into account random failure processes for the system components, the classical concept of probability is commonly used to quantify system reliability under uncertainty.

These basic concepts have led to much theory and many successful applications, for example on system design, inspection and maintenance, and general risk assessment, for a wide variety of systems and networks. In recent years, attention has spread from core reliability theory to the wider concept of system dependability [18]. This encompasses a variety of related concepts in addition

W. Zamojski and J. Sugier (eds.), *Dependability Problems of Complex Information Systems*, Advances in Intelligent Systems and Computing 307, DOI: 10.1007/978-3-319-08964-5_2

to reliability, such as availability, maintainability, safety, security, flexibility, resilience and integrity of the system and its functioning. While all these have intuitively clear meanings, the literature has provided different definitions and interpretations for each, often related to varying application areas, circumstances and requirements. This wider view of dependability is particularly important when real-world scenarios are considered, as classical reliability theory is often based on assumptions made for theoretical convenience but not always justified in applications.

In this paper, we explore some uncertain or unknown aspects related to a system's functioning, and we suggest a simple way for taking these into account in quantification of reliability of a system. The main idea is that the system may have to deal with a variety of tasks of different types, which put different requirements on the system. We focus then on a specific future task to be performed, calling it the 'next task', and take uncertainty about the type of this task into account by using probabilities over the different types of tasks, and by generalizing this to imprecise probabilities. This enables uncertainty and indeterminacy to be included in the modelling. This approach is very flexible, it can even be used to include the possibility of a fully unknown type of task, which might for example be suitable to reflect possible unknown threats to the system.

Section 2 presents the structure function as a, possibly imprecise, probability, the corresponding use in (lower and upper) survival signatures is presented in Section 3. The uncertainty with regard to the type of the next task is considered in Section 4 and illustrated via an example in Section 5. The paper concludes with a discussion of some related aspects in Section 6, which suggests several ways in which the concepts proposed in this paper can be used for uncertainty quantification of aspects of system dependability. The main aim of this paper is to trigger further research using the flexibility provided by the (lower and upper) survival signatures.

2 The Structure Function as (Imprecise) Probability

The first proposal presented and discussed here is to generalize the structure function to reflect uncertainty about the system's functioning given the state vector $\underline{x}$, by defining it as a probability, so $\phi : \{0,1\}^m \to [0,1]$. We define $\phi(\underline{x})$ as the probability that the system functions for a specific state vector $\underline{x}$ and for the next task the system is required to perform. Let S denote the event that the system functions as required for the next task it is demanded to perform, then

$$\phi(\underline{x}) = P(S|\underline{x}) \tag{1}$$

We have kept the same notation for the structure function, as a probability, as in Section 1, which should not cause problems and is justified as the earlier definition of structure function can be regarded as a special case of this generalized definition with all probabilities either 0 or 1. We should emphasize that we consider system functioning explicitly for the next task that the system has to perform, which varies from the usual definition for system functioning in the literature. We do so as the generalization considered in this paper is particularly aimed at dealing with different types of tasks, which is easiest when focussing explicitly on the next task; we discuss this in more detail later. This can, quite straightforwardly, be generalized to considering multiple future tasks, we do not discuss this further in this paper.

This generalization already enables an important range of real-world scenarios to be modelled in a straightforward way. Furthermore, as we will discuss in Section 3, it can quite easily be embedded in existing theory for reliability quantification. Scenarios where the flexibility of the structure function as a probability might be useful are, of course, situations where even with known status of the components, it is not certain whether or not the system functions, that is performs its task as required. This may be due to varying circumstances or requirements which may not be modelled explicitly, or may not even be fully known. For example, one could consider a wind farm, a collection of wind turbines at a specific location, as one system, with the task to generate a level of energy required to provide a specific area with sufficient electricity. One could consider each wind turbine as a component (with several other types of components in the system, that is irrelevant for now). Even if one knows the number of functioning components at a particular time, factors such as the weather, the availability of other electricity generating resources for the network, and the specific electricity demand, can lead to uncertainty about whether or not the system meets the actual requirements. To fit with the established deterministic definition of the structure function one can define system functioning in far more detail, but this may be hard to do in practice. As another example, one could think about a network of computers which together form a system for complex computations, where its actual success in dealing with required tasks might be achieved with some computers not functioning, but with some lack of knowledge about the exact number of computers required to complete tasks of different types.

The generalization to consider the structure function as a probability, although mathematically straightforward, requires substantial information in order to assess the probabilities of system functioning for all possible state vectors $\underline{x}$. While this modelling might explicitly take co-variates into account, thus possibly benefitting from a large variety of statistical models, it may be difficult to actually formulate the important co-variates and one might not know their specific values. This leads to two further topics we wish to discuss, namely what precisely is meant when we say that the system functions, and a generalization of probability to allow lack of knowledge to be reflected.

Whether or not a real-world system performs its task well may depend on many circumstances beyond the states of the system components. It may be too daunting to specify system functioning for all possible circumstances, and it may even be impossible to know all possible circumstances. Hence, speaking of 'system functioning' in the traditional theoretic way seems rather restricted. One suggestion would be to only define system functioning for one (or a specified number of) application(s), e.g. whether or not a system functions at its next required use. This will not be sufficient for all real-world scenarios, but it will enable important aspects of uncertainty on factors such as different tasks and circumstances to be taken into account. We believe that this is a topic that requires further attention, it links to many system dependability concepts including flexibility and resilience.

The generalization of the structure function as a probability provides substantial enhanced modelling opportunities for system dependability. However, the concept of probability, while being well established and very successfully applied in most areas of human activity involving uncertainty, is not sufficiently flexible to quantify and reflect the multi-dimensional nature of uncertainty. In particular, the use of single-valued probabilities for events does not enable the strength or lack of information to be taken into account, with most obvious limitation the inability to reflect if 'no information at all' is available about an event of interest. In recent decades, theory of imprecise probability [3,11] has gained increasing attention from the research community, including contributions to reliability and risk [12]. It generalizes classical, precise, probability theory by assigning to each event two values, a lower probability and an upper probability, denoted by $\underline{P}$ and $\overline{P}$, respectively, with $0 \leq \underline{P} \leq \overline{P} \leq 1$. These can be interpreted in several ways [3,11], for the current discussion it suffices to regard them as the sharpest bounds for a probability based on the information available, where the lower probability typically reflects the information available in support of the event of interest and the corresponding upper probability reflects the information available against this event. The case of no information at all can be reflected by $[\underline{P}, \overline{P}] = [0, 1]$ while equality $\underline{P} = \overline{P}$ results in classical precise probability.

We propose the further generalization of the structure function within imprecise probability theory by introducing the *lower structure function*

$$\underline{\phi}(\underline{x}) = \underline{P}(S|\underline{x}) \tag{2}$$

and the *upper structure function*

$$\overline{\phi}(\underline{x}) = \overline{P}(S|\underline{x}) \tag{3}$$

This provides substantial flexibility for practical application of methods to quantify system reliability and other dependability concepts. For example, it may be known historically that, under different external circumstances, a system with a certain subset of its components functioning manages a task well in 85 to 95 percent of all cases. While it might be possible to go into further detail and e.g. describe beliefs within this range by a probability distribution, or assume this for mathematical convenience, this may not be required or it may actually be impossible in a meaningful way, and one can use lower probability 0.85 and upper probability 0.95 to accurately reflect this information. If one has to rely on expert judgements to assign the values of the structure function, then time may often be too limited to meaningfully assign precise probabilities for system functioning for all possible component state vectors. In such cases, the use of imprecise probabilities also offers suitable flexibility. Assigning a subset of probabilities for some events (or bounds for these) will imply bounds for all other related events under suitable coherence assumptions [3,11][1], where particularly assumed coherence of the system, which implies that any additional component failure can never improve system functioning, is useful and practically justifiable in many applications.

3 Survival Signature with Generalized Structure Function

Recently, we introduced the survival signature to assist reliability analyses for systems with multiple types of components [9]. In case of just a single type of components, the survival signature is closely related to the system signature [17], which is well-established and the topic of many research papers during the last decade. However, generalization of the signature to systems with multiple types of components is extremely complicated (as it involves ordering order statistics of different distributions), so much so that it cannot be applied to most practical systems. In addition to the possible use for such systems, where the benefit only occurs if there are multiple components of the same types, the survival signature is arguably also easier to interpret than the signature. In this section, we briefly review the survival signature and some recent advances, then link it to the generalization of the structure function proposed in Section 2.

Consider a system with $K \geq 1$ types of components, with m_k components of type $k \in \{1, \ldots, K\}$ and $\sum_{k=1}^{K} m_k = m$. Assume that the random failure times of components of the same type are exchangeable [14], while full independence is assumed for the random failure times of components of different types. Due to the arbitrary ordering of the components in the state vector, components of the same type can be grouped together, leading to a state vector that can be written as $\underline{x} = (\underline{x}^1, \underline{x}^2, \ldots, \underline{x}^K)$, with $\underline{x}^k = (x_1^k, x_2^k, \ldots, x_{m_k}^k)$ the sub-vector representing the states of the components of type k.

The survival signature [9] for such a system, denoted by $\Phi(l_1, \ldots, l_K)$, with $l_k = 0, 1, \ldots, m_k$ for $k = 1, \ldots, K$, is defined as the probability for the event that the system functions given that *precisely* l_k of its m_k components of type k function, for each $k \in \{1, \ldots, K\}$.

There are $\binom{m_k}{l_k}$ state vectors $\underline{x}^k$ with $\sum_{i=1}^{m_k} x_i^k = l_k$. Let $S_{l_k}^k$ denote the set of these state vectors for components of type k and let $S_{l_1,\ldots,l_K}$ denote the set of all state vectors for the whole system for which $\sum_{i=1}^{m_k} x_i^k = l_k$, $k = 1, \ldots, K$. Due to the exchangeability assumption for the failure times of the m_k components of type k, all the state vectors $\underline{x}^k \in S_{l_k}^k$ are equally likely to occur, hence [9]

$$\Phi(l_1, \ldots, l_K) = \left[\prod_{k=1}^{K} \binom{m_k}{l_k}^{-1} \right] \times \sum_{\underline{x} \in S_{l_1,\ldots,l_K}} \phi(\underline{x}) \tag{4}$$

[1] Coherence here refers to consistency properties of imprecise probabilities, so is different from the term 'coherence' used for systems; we do not use this term in the former meaning further in this paper to avoid confusion.

We now consider the survival signature with the generalized structure function as discussed in Section 2, using the lower structure function (2) and upper structure function (3). The survival signature can straightforwardly be adapted to include these, due to its monotone dependence on the structure function. This leads to the following definitions of the *lower survival signature*

$$\underline{\Phi}(l_1,\ldots,l_K) = \left[\prod_{k=1}^{K} \binom{m_k}{l_k}^{-1}\right] \times \sum_{\underline{x} \in S_{l_1,\ldots,l_K}} \underline{\phi}(\underline{x}) \tag{5}$$

and the corresponding *upper survival signature*

$$\overline{\Phi}(l_1,\ldots,l_K) = \left[\prod_{k=1}^{K} \binom{m_k}{l_k}^{-1}\right] \times \sum_{\underline{x} \in S_{l_1,\ldots,l_K}} \overline{\phi}(\underline{x}) \tag{6}$$

These are the sharpest possible bounds for the survival signature corresponding to the lower and upper structure functions, and as such indeed the lower and upper probabilities for the event that the system functions given that precisely l_k of its m_k components of type k function, for each $k \in \{1,\ldots,K\}$.

These lower and upper survival signatures can be used for imprecise reliability quantifications. Particularly if chosen quantifications are monotone functions of the survival signature, this is again a straightforward generalization of the precise approach [9]. Let us consider the event that the system functions for the next task it has to perform, denoted by S. Let $C_k \in \{0,1,\ldots,m_k\}$ denote the number of components of type k in the system which function when required for the next task. The probability for the event S is [9]

$$P(S) = \sum_{l_1=0}^{m_1} \cdots \sum_{l_K=0}^{m_K} \Phi(l_1,\ldots,l_K) P(\bigcap_{k=1}^{K} \{C_k = l_k\}) \tag{7}$$

With the generalization of the survival signature, we get the lower probability for the event that the systems functions for the next task

$$\underline{P}(S) = \sum_{l_1=0}^{m_1} \cdots \sum_{l_K=0}^{m_K} \underline{\Phi}(l_1,\ldots,l_K) P(\bigcap_{k=1}^{K} \{C_k = l_k\}) \tag{8}$$

and the corresponding upper probability

$$\overline{P}(S) = \sum_{l_1=0}^{m_1} \cdots \sum_{l_K=0}^{m_K} \overline{\Phi}(l_1,\ldots,l_K) P(\bigcap_{k=1}^{K} \{C_k = l_k\}) \tag{9}$$

For this imprecise case, just as for the precise case [9], assuming independence of the functioning of components of different types leads to, for $l_k \in \{0,1,\ldots,m_k\}$ for each $k \in \{1,\ldots,K\}$,

$$P(\bigcap_{k=1}^{K} \{C_k = l_k\}) = \prod_{k=1}^{K} P(C_k = l_k)$$

If in addition it is assumed that functioning of components of the same type is conditionally independent given probability $f_k \in [0,1]$ that a component of type k functions for the next task, then

$$P(\bigcap_{k=1}^{K}\{C_k = l_k\}) = \prod_{k=1}^{K} \binom{m_k}{l_k} f_k^{l_k} [1 - f_k]^{m_k - l_k}$$

This leads to relatively straightforward computations for reliability metrics, which we do not discuss further in this paper. It is important though to emphasize that exactly the same approach can be followed when interest is in processes over time, where instead of focussing on functioning of the system for the next task one can consider the probability that the system functions at a given time [9].

The probabilities for the numbers of functioning components can also be generalized to lower and upper probabilities, as e.g. done by Coolen et al. [10] within the nonparametric predictive inference framework of statistics [5], where lower and upper probabilities for the events $C_k = l_k$ are inferred from test data on components of the same types as those in the system. This step is slightly less trivial as one must ensure to have probability distributions for these events, thus summing to one over $l_k = 0, 1, \ldots, m_k$ for each type k. For monotone systems this is not very complicated due to the monotonicity of the (lower or upper) survival signature.

The main advantage of the survival signature, in line with this property of the signature for systems with a single type of components [17], as shown by Equation (7), is that the information about the system structure is fully separated from the information about functioning of the components, which simplifies related statistical inference as well as considerations of optimal system design. This property clearly also holds for the lower and upper survival signatures as is shown by Equations (8) and (9).

4 Multiple Types of Tasks

If a system may need to deal with different tasks, the (lower or upper) structure function should, ideally, be defined for each specific type of task. Let there be $R \geq 1$ types of tasks. The (lower or upper) structure function for a specific type of task $r \in \{1, \ldots, R\}$ is the (lower or upper) probability for the event that the system functions for component states $\underline{x}$ and for known type of task r, we denote these as before with an additional subscript r (we generalize earlier notation in this way throughout this section without explicit introduction), so

$$\phi_r(\underline{x}) = P(S|\underline{x}, r) \qquad \underline{\phi}_r(\underline{x}) = \underline{P}(S|\underline{x}, r) \qquad \overline{\phi}_r(\underline{x}) = \overline{P}(S|\underline{x}, r)$$

If interest is in the next task that the system has to perform, and it is known of which type this task is, then we are back to the setting discussed before. If the type of task is not known with certainty, then there are several possible scenarios. First, suppose that one can assign a precise probability for the event that the next task is of type r, denoted by p_r, for each $r \in \{1, \ldots, R\}$. Then the system structure function for the next task can be derived via the theorem of total probability, which also applies straightforwardly to the corresponding lower and upper structure functions in the generalized case. This leads to

$$\phi(\underline{x}) = \sum_{r=1}^{R} \phi_r(\underline{x}) p_r \qquad \underline{\phi}(\underline{x}) = \sum_{r=1}^{R} \underline{\phi}_r(\underline{x}) p_r \qquad \overline{\phi}(\underline{x}) = \sum_{r=1}^{R} \overline{\phi}_r(\underline{x}) p_r$$

For this scenario the corresponding lower and upper survival signatures that apply for the next task, of random type, are easily derived and given by

$$\begin{aligned}\underline{\Phi}(l_1, \ldots, l_K) &= \left[\prod_{k=1}^{K} \binom{m_k}{l_k}^{-1} \right] \times \sum_{\underline{x} \in S_{l_1, \ldots, l_K}} \sum_{r=1}^{R} \underline{\phi}_r(\underline{x}) p_r \\ &= \sum_{r=1}^{R} \underline{\Phi}_r(l_1, \ldots, l_K) p_r \\ \overline{\Phi}(l_1, \ldots, l_K) &= \left[\prod_{k=1}^{K} \binom{m_k}{l_k}^{-1} \right] \times \sum_{\underline{x} \in S_{l_1, \ldots, l_K}} \sum_{r=1}^{R} \overline{\phi}_r(\underline{x}) p_r \\ &= \sum_{r=1}^{R} \overline{\Phi}_r(l_1, \ldots, l_K) p_r\end{aligned}$$

These results hold as all sums involved are finite, hence the order of summations can be changed, which can also be applied to derive

$$\underline{P}(S) = \sum_{r=1}^{R} \underline{P}_r(S) p_r$$

$$\overline{P}(S) = \sum_{r=1}^{R} \overline{P}_r(S) p_r$$

Secondly, one may only be able to assign bounds for the probabilities p_r, where the sharpest bounds one can assign are lower and upper probabilities, denoted by $\underline{p}_r$ and $\overline{p}_r$. Let p denote any probability vector of dimension R, so $p = (p_1, \ldots, p_R)$ with all $p_r \geq 0$ and $\sum_{r=1}^{R} p_r = 1$, and let $\mathcal{P}$ denote the set of all such probability vectors with $\underline{p}_r \leq p_r \leq \overline{p}_r$ for all $r \in \{1, \ldots, R\}$[2]. In this situation, deriving the lower and upper structure functions for the next task is less straigthforward, as they require optimisation over the set $\mathcal{P}$ of probability vectors

$$\underline{\phi}(\underline{x}) = \min_{p \in \mathcal{P}} \sum_{r=1}^{R} \underline{\phi}_r(\underline{x}) p_r \qquad \overline{\phi}(\underline{x}) = \max_{p \in \mathcal{P}} \sum_{r=1}^{R} \overline{\phi}_r(\underline{x}) p_r \tag{10}$$

[2] This set $\mathcal{P}$ is known as the 'structure' of the imprecise probability model [3,11], we will not use this term further to avoid confusion with the use of the term structure for the considered system.

In case of a precise structure function, the lower and upper structure functions on the right-hand sides of these equations are just equal to the precise structure function, with imprecision still resulting from the set $\mathcal{P}$ of probability vectors. While these optima are not available in closed-form, their computation is quite straightforward, solutions are obtained by setting all p_r equal to either $\underline{p}_r$ or $\overline{p}_r$ apart from one which will take on a value within its corresponding range $[\underline{p}_r, \overline{p}_r]$ such that the individual probabilities sum up to one.

For this scenario, deriving the corresponding lower and upper survival signatures is less straightforward than for the first scenario above. Inserting the lower and upper structure functions (10) into the equations for the lower and upper survival signatures would give the expressions

$$\left[\prod_{k=1}^{K}\binom{m_k}{l_k}^{-1}\right] \times \sum_{\underline{x}\in S_{l_1,\ldots,l_K}} \left(\min_{p\in\mathcal{P}} \sum_{r=1}^{R} \underline{\phi}_r(\underline{x})p_r\right) \tag{11}$$

and

$$\left[\prod_{k=1}^{K}\binom{m_k}{l_k}^{-1}\right] \times \sum_{\underline{x}\in S_{l_1,\ldots,l_K}} \left(\max_{p\in\mathcal{P}} \sum_{r=1}^{R} \overline{\phi}_r(\underline{x})p_r\right) \tag{12}$$

However, the corresponding lower and upper survival signatures are

$$\underline{\Phi}(l_1,\ldots,l_K) = \min_{p\in\mathcal{P}}\left(\left[\prod_{k=1}^{K}\binom{m_k}{l_k}^{-1}\right] \times \sum_{\underline{x}\in S_{l_1,\ldots,l_K}} \sum_{r=1}^{R} \underline{\phi}_r(\underline{x})p_r\right)$$
$$\overline{\Phi}(l_1,\ldots,l_K) = \max_{p\in\mathcal{P}}\left(\left[\prod_{k=1}^{K}\binom{m_k}{l_k}^{-1}\right] \times \sum_{\underline{x}\in S_{l_1,\ldots,l_K}} \sum_{r=1}^{R} \overline{\phi}_r(\underline{x})p_r\right)$$

which generally requires solving complex optimisation problems. This lower survival signature is greater than or equal to expression (11) and this upper survival signature is less than or equal to expression (12). If the optimisations in expression (11) all have the same probability vector within $\mathcal{P}$ as solution, then the lower survival signature is equal to this expression, and similarly for the upper survival signature with regard to the optimisations in expression (12). While this may appear to be unlikely, we will illustrate a case were it applies in the example in Section 5. Further investigations into the optimisation problems for general situations are left as an important challenge for future research.

Finally, one may wish to use statistical inference for the p_r in case one has relevant data. There is a variety of options, including Bayesian methods, which might be generalized through the use of sets of prior distributions as in the imprecise Dirichlet model for multinomial data [3] and nonparametric predictive inference [7,8]. The latter approach may be of specific interest as it provides the possibility to take unobserved or even undefined tasks into consideration [4].

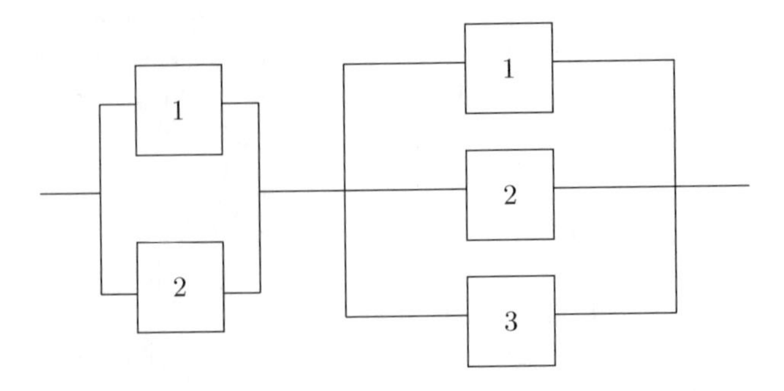

Fig. 1. System with three types of components

Table 1. Survival signatures for system in Figure 1, two cases

l_1	l_2	l_3	$\Phi_1(l_1,l_2,l_3)$	$\Phi_2(l_1,l_2,l_3)$
0	1	1	1/2	0
0	2	0	1	0
1	0	1	1/2	0
1	1	0	1/2	0
1	1	1	3/4	1/2
1	2	0	1	1/2
2	0	0	1	0
2	1	0	1	1/2

5 Example

Consider the system presented in Figure 1, consisting of two subsystems in series configuration, but with the following variation for the second subsystem consisting of three components: for some tasks to be performed according to the requirements it is sufficient for one of the three components to function, but for other tasks (or under other circumstances) it is necessary to have at least two components functioning. We will refer to these as Case 1 and Case 2, respectively. The survival signatures for this system corresponding to these two cases are presented in Table 1, denoted by Φ_1 and Φ_2, where the quite trivial entries for which both survival signatures are equal to 0 or 1 are not included.

Suppose that five different possible tasks have been identified which this system may have to deal with. This may actually be different tasks, or just due to different circumstances under which the tasks may need to be performed. For Task A Case 1 applies, so only one functioning component in the second subsystem is required. For Task B Case 2 applies. For Task C there is uncertainty about whether one or two components need to function in the second subsystem, with either case having probability 1/2. For Task D the same uncertainty occurs, but the probabilities that either case applies are not precisely known, with lower and upper probability for Case 1 equal to 0.4 and 0.8, respectively, which by the conjugacy property for lower and upper probabilities [11] implies

Table 2. Lower and upper survival signatures for Tasks A-E

l_1	l_2	l_3	Φ_A	Φ_B	Φ_C	$[\underline{\Phi}_D, \overline{\Phi}_D]$	$[\underline{\Phi}_E, \overline{\Phi}_E]$
0	1	1	0.5	0	0.25	$[0.2, 0.4]$	$[0, 0.5]$
0	2	0	1	0	0.5	$[0.4, 0.8]$	$[0, 1]$
1	0	1	0.5	0	0.25	$[0.2, 0.4]$	$[0, 0.5]$
1	1	0	0.5	0	0.25	$[0.2, 0.4]$	$[0, 0.5]$
1	1	1	0.75	0.5	0.625	$[0.6, 0.7]$	$[0.5, 0.75]$
1	2	0	1	0.5	0.75	$[0.7, 0.9]$	$[0.5, 1]$
2	0	0	1	0	0.5	$[0.4, 0.8]$	$[0, 1]$
2	1	0	1	0.5	0.75	$[0.7, 0.9]$	$[0.5, 1]$

lower and upper probability 0.2 and 0.6 for Case 2. Finally, for Task E the same uncertainty occurs but there is no knowledge at all about the probability with which each case applies, represented by lower and upper probabilities 0 and 1, respectively, for both cases.

The survival signatures for Tasks A and B are just $\Phi_A = \Phi_1$ and $\Phi_B = \Phi_2$. For Tasks C-E, the generalized structure functions are easily derived and lead to the (lower and upper) survival signatures given in Table 2, where for completeness also Φ_A and Φ_B are given and entries which are either equal to 0 or 1 for all these functions have been left out.

For these (lower and upper) survival signatures, the following ordering holds for all (l_1, l_2, l_3),

$$\Phi_B = \underline{\Phi}_E \leq \underline{\Phi}_D \leq \Phi_C \leq \overline{\Phi}_D \leq \overline{\Phi}_E = \Phi_A$$

This means that in this example the special case applies in which expressions (11) and (12) give the lower and upper survival signatures, as the minimisations to derive the following lower survival signatures are all solved by the same probability vector in $\mathcal{P}$, and similar for the maximisations to derive the upper survival signatures. While this special case does not illustrate the full modelling ability of the concepts presented in this paper, it is of practical interest in scenarios such as discussed in this example, where there are a number of basic tasks which differ with regard to their demands on the system, and a variety of cases for the next possible task to be performed, each of these being represented by a different (imprecise) probability distribution over those basic tasks. For all such cases, the optimisations involved in deriving the lower and upper survival signatures for the next task to be performed by the system are straightforward, as in this example. We now consider several scenarios with different levels of knowledge about the type of the next task, the lower and upper survival signatures are presented in Table 3 (again leaving out those which are trivially equal to 0 or 1).

Suppose first, Case I, that the next task can be of any of the five types $A-E$, each with probability 0.2. The lower survival signature for the next task in this

Table 3. Lower and upper survival signatures for Cases *I*-*IV*

l_1 l_2 l_3	$[\underline{\Phi}_I, \overline{\Phi}_I]$	$[\underline{\Phi}_{II}, \overline{\Phi}_{II}]$	$[\underline{\Phi}_{III}, \overline{\Phi}_{III}]$	$[\underline{\Phi}_{IV}, \overline{\Phi}_{IV}]$
0 1 1	$[0.19, 0.33]$	$[0.17, 0.34]$	$[0.095, 0.415]$	$[0.17, 0.39]$
0 2 0	$[0.38, 0.66]$	$[0.34, 0.68]$	$[0.19, 0.83]$	$[0.34, 0.68]$
1 0 1	$[0.19, 0.33]$	$[0.17, 0.34]$	$[0.095, 0.415]$	$[0.17, 0.39]$
1 1 0	$[0.19, 0.33]$	$[0.17, 0.34]$	$[0.095, 0.415]$	$[0.17, 0.39]$
1 1 1	$[0.595, 0.665]$	$[0.585, 0.67]$	$[0.5475, 0.7075]$	$[0.535, 0.695]$
1 2 0	$[0.69, 0.83]$	$[0.67, 0.84]$	$[0.595, 0.915]$	$[0.62, 0.84]$
2 0 0	$[0.38, 0.66]$	$[0.34, 0.68]$	$[0.19, 0.83]$	$[0.34, 0.68]$
2 1 0	$[0.69, 0.83]$	$[0.67, 0.84]$	$[0.595, 0.915]$	$[0.62, 0.84]$

case, denoted by $\underline{\Phi}_I$, is derived as the average of the (lower) survival signatures for tasks A-E, and similar for the upper survival signature. For Case II, suppose that the next task can again be of types A, B or C with probability 0.2 each, but there is uncertainty ('indeterminacy') with regard to the probability that this task may be of types D or E, reflected through lower and upper probabilities of 0.1 and 0.3, respectively, for both these types. To derive the lower survival signature for the next task in this case, we assign maximum probability 0.3 to $\underline{\Phi}_E$ for all (l_1, l_2, l_3), as this is never greater than $\underline{\Phi}_D$, which of course is assigned the minimum possible probability 0.1 to remain within the set of probability vectors $\mathcal{P}$. Similarly, due to $\overline{\Phi}_E \geq \overline{\Phi}_D$ for all (l_1, l_2, l_3), the corresponding upper survival signature is derived by assigning probability 0.3 to $\overline{\Phi}_E$ and 0.1 to $\overline{\Phi}_D$.

To illustrate a greater level of indeterminacy with regard to the next task, Case III considers that it may be of each of the five identified types with lower probability 0.1 and upper probability 0.5. With the ordering of the (lower and upper) survival signatures for the five types, it is easy to verify that the lower survival signature over this set of probability vectors $\mathcal{P}$ is derived by assigning probability 0.4 to Φ_B, 0.3 to $\underline{\Phi}_E$ and 0.1 to each of $\underline{\Phi}_D$, Φ_C and Φ_A. Similarly, the upper survival signature is derived by assigning probability 0.4 to Φ_A, 0.3 to $\overline{\Phi}_E$ and 0.1 to each of $\overline{\Phi}_D$, Φ_C and Φ_B.

Finally, we return to the scenario of Case II, but with an important addition. For Case IV, suppose that it is judged that the next task the system needs to perform could actually also be a totally unknown task, for which it is not known at all whether or not the system can deal with it. This goes beyond the two basic tasks discussed throughout this example, for which the structure functions were given in Table 1. To reflect total lack of knowledge of such an unknown ('unidentified', 'unforeseen') task, which we indicate by index U, we can assign lower structure function $\underline{\phi}_U(l_1, l_2, l_3) = 0$ and upper structure function $\overline{\phi}_U(l_1, l_2, l_3) = 1$ for all (l_1, l_2, l_3), reflecting that even with all components functioning we do not know if the system can deal with this task, and that even with no components functioning it might be possible that this task can be satisfactorily dealt with. While these values may appear to be extreme, it covers all possibilities for unknown tasks, including e.g. targeted attacks on the system. It should be emphasized that such lack of knowledge cannot be taken into account adequately when restricted to the use of precise probabilities, and thus illustrates one of the major advantages of the use of imprecise probabilities. Let us assume that the next task can be of type U with lower probability 0 and upper probability 0.1, so the set of probability vectors over the six types $A - E$ and U consists of all probability vectors with $p_A = p_B = p_C = 0.2$, $p_D, p_E \in [0.1, 0.3]$ and $p_U \in [0, 0.1]$. To derive the lower survival signature for the next task in this case, we assign, in addition to the fixed probabilities 0.2 to types A, B, C, probability 0.1 to $\underline{\Phi}_U$, 0.2 to $\underline{\Phi}_E$ and 0.1 to $\underline{\Phi}_D$. To derive the corresponding upper survival signature, we similarly assign probability 0.1 to $\overline{\Phi}_U$, 0.2 to $\overline{\Phi}_E$ and 0.1 to $\overline{\Phi}_D$.

As is clear from Table 3, increase in indeterminacy, reflected through increased imprecision in the assigned lower and upper probabilities, leads to more imprecise lower and upper survival signatures in a logically nested way. From the perspective of risk management, the lower survival signatures are likely to be of most interest, as they reflect the most pessimistic scenario for system functioning corresponding to the information and assumptions made. As this example shows, the lower survival signature is derived by assigning the maximum possible probabilities to the possible types of task for which the system is least likely to function well.

In Case IV, we illustrated the possibility to include a totally unknown type of task by assigning lower and upper probabilities of 0 and 0.1 for the event that the next task is of such nature. In most risk scenarios, it would make sense to have lower probability 0 for such an event. The upper probability is, of course, more important for risk management as, combined with the lower probability for the system functioning well for such a task, it relates to the most pessimistic scenario. To illustrate our method we just chose the value 0.1 for this upper probability, yet it is worth mentioning that the nonparametric predictive inference (NPI) approach can actually provide a meaningful numerical value for the upper probability for the event that an as yet unobserved or even undefined event occurs [4,7,8,13]. This NPI upper probability, which we do not discuss further in this paper, is based on relatively weak assumptions and is decreasing as function of the number of events considered in the data yet increasing as function of the number of different types of tasks the system had to deal with thus far.

6 Discussion

Traditional theory of system reliability tends to make some pretty strong assumptions with regard to knowledge about systems and their practical use. As shown in this paper, rather straightforward generalization of the structure function to consider it as a probability increases modelling opportunities substantially. Beyond that, the use of imprecise probabilities enables us to reflect indeterminacy, which is particularly important in risk scenarios where one may have limited knowledge and experience of the system functioning, or where the system may need to be resilient in case of unforeseen tasks. In this paper we have illustrated the approach mainly by considering different types of tasks, which in the example were related to two basic ways a given system could need to function, namely with one subsystem either requiring only one or at least two of its three components to function. The main advantage of the survival signature as presented in this paper is that this generalization of the structure function is quite straigthforwardly embedded in its definition, leading to lower and upper survival signatures. These are formulated for a single future task, which is important if one wishes to use statistical methods to infer system reliability and to reflect the amount of information available. Developing such statistical methods related to the lower and upper survival signatures is an interesting challenge for future research.

One could argue that using imprecise probability to reflect indeterminacy is an easy way out, as one effectively considers both the most optimistic and pessimistic scenarios which correspond to the information available, and reports the bounds based on these as the results of the inferences. The importance of this generalization of probability should, however, not be underestimated, as it avoids choosing precise values even in cases where there is no justification for doing so. Seeing the quality of the available information reflected explicitly in the reliability quantification, without lack of detailed information being hidden due to stronger assumptions or precise input values chosen for convenience, provides useful information for managing risks. If one does have quite detailed information it can be included in the inferences, and indeed doing so will normally lead to less imprecision, so it is certainly worth aiming to use all available information. In addition, one can also explore the influence of further assumptions or information on the imprecise results, which can be helpful if one wishes to explore what to focus on in order to derive the most useful information for a specific problem.

Following the first steps presented in this paper, there are many research challenges in order to develop a methodology that is applicable to large scale systems. It is important for such research challenges to be taken on with direct relation to real world applications, in order to discover the real problems and to see how results can be implemented. Part of such challenges will be in computation, as deriving the survival signature involves complex calculations, the number of which increases exponentially with the size of the system. Aslett [2] has developed a function in the statistical software R which can compute the survival signature for small to medium sized systems, but for practical systems and networks more research is required.

The theory presented in this paper is particularly useful for systems and networks with multiple types of components and with many components of the same type, as the survival signature is a sufficient summary of the system's structure which, in such cases, provides a substantial reduction compared to the complete structure function. One might encounter such systems and networks in many application areas, for example complex computer or communication systems with many parallel servers, energy networks, and transport infrastructure including rail networks. It may further be relevant for biology and medical research, exploring the opportunities for applications is an exciting challenge. In many modern applications emphasis is on real-time monitoring and online prediction [16]. The setting presented in this paper may be suitable for such inference, in particular when combined with nonparametric predictive inference (NPI) [5,9] where inferences are in terms of the next event and take all data into account. The combined use of NPI and signatures has been presented for systems consisting of only a single type of components [1,6]. Recently, NPI has also been applied together with the survival signature [10], this also requires a substantial research effort to become implementable to large scale practical problems.

References

1. Al-nefaiee, A.H., Coolen, F.P.A.: Nonparametric predictive inference for system failure time based on bounds for the signature. Journal of Risk and Reliability 227, 513–522 (2013)
2. Aslett, L.J.M.: ReliabilityTheory: Tools for structural reliability analysis. R package (2012), `www.louisaslett.com`
3. Augustin, T., Coolen, F.P.A., de Cooman, G., Troffaes, M.C.M.: Introduction to Imprecise Probabilities. Wiley, Chichester (2014)
4. Coolen, F.P.A.: Nonparametric prediction of unobserved failure modes. Journal of Risk and Reliability 221, 207–216 (2007)
5. Coolen, F.P.A.: Nonparametric predictive inference. In: Lovric (ed.) International Encyclopedia of Statistical Science, pp. 968–970. Springer (2011), `www.npi-statistics.com`
6. Coolen, F.P.A., Al-nefaiee, A.H.: Nonparametric predictive inference for failure times of systems with exchangeable components. Journal of Risk and Reliability 226, 262–273 (2012)
7. Coolen, F.P.A., Augustin, T.: Learning from multinomial data: a nonparametric predictive alternative to the Imprecise Dirichlet Model. In: Cozman, et al. (eds.) Proceedings ISIPTA 2005, pp. 125–134 (2005)
8. Coolen, F.P.A., Augustin, T.: A nonparametric predictive alternative to the Imprecise Dirichlet Model: the case of a known number of categories. International Journal of Approximate Reasoning 50, 217–230 (2009)
9. Coolen, F.P.A., Coolen-Maturi, T.: Generalizing the signature to systems with multiple types of components. In: Zamojski, W., Mazurkiewicz, J., Sugier, J., Walkowiak, T., Kacprzyk, J. (eds.) Complex Systems and Dependability. Advances in Intelligent Systems and Computing, vol. 170, pp. 115–130. Springer, Heidelberg (2012)
10. Coolen, F.P.A., Coolen-Maturi, T., Al-nefaiee, A.H.: Nonparametric predictive inference for system reliability using the survival signature. Journal of Risk and Reliability, doi:10.1177/1748006X14526390
11. Coolen, F.P.A., Troffaes, M.C.M., Augustin, T.: Imprecise probability. In: Lovric (ed.) International Encyclopedia of Statistical Science, pp. 645–648. Springer (2011)
12. Coolen, F.P.A., Utkin, L.V.: Imprecise reliability. In: Lovric (ed.) International Encyclopedia of Statistical Science, pp. 649–650. Springer (2011)
13. Coolen-Maturi, T., Coolen, F.P.A.: Unobserved, re-defined, unknown or removed failure modes in competing risks. Journal of Risk and Reliability 225, 461–474 (2011)

14. De Finetti, B.: Theory of Probability. Wiley, New York (1974)
15. Maturi, T.A., Coolen-Schrijner, P., Coolen, F.P.A.: Nonparametric predictive inference for competing risks. Journal of Risk and Reliability 224, 11–26 (2010)
16. Salfner, F., Lenk, M., Malek, M.: A survey of online failure prediction methods. ACM Computing Surveys 42(3), article 10 (2010)
17. Samaniego, F.J.: System Signatures and their Applications in Engineering Reliability. Springer (2007)
18. Zamojski, W., Mazurkiewicz, J.: From reliability to system dependability - theory and models. In: Kolowrocki, Soszynska-Budny (eds.) Proceedings SSARS 2011 - 5th Summer Safety & Reliability Seminars, vol. 1, pp. 223–231 (2011)

Improving the Dependability of Distributed Surveillance Systems Using Diverse Redundant Detectors

Francesco Flammini[1], Nicola Mazzocca[2], Alfio Pappalardo[1,2], Concetta Pragliola[1], and Valeria Vittorini[2]

[1] Ansaldo STS, Innovation & Competitiveness Unit
via Argine 425, Naples, Italy
{francesco.flammini,alfio.pappalardo, concetta.pragliola}@ansaldo-sts.com

[2] University of Naples "Federico II", Department of Computer & Systems Engineering, via Claudio 21, Naples, Italy
{nicola.mazzocca,alfio.pappalardo,valeria.vittorini}@unina.it

Abstract. Sensor networks nowadays employed in critical monitoring and surveillance applications represent a relevant case of complex information infrastructures whose dependability needs to be carefully assessed. Detection models based on Event Trees provide a simple and effective mean to correlate events in Physical Security Information Management (PSIM) systems. However, as a deterministic modeling approach, Event Trees are not able to address uncertainties in practical applications, like: 1) imperfect threat modelling; 2) sensor false alarms. Regarding point (1), it is quite obvious that real-world threat scenarios can be very variable and it is nearly impossible to consider all the possible combinations of events characterizing a threat. Point (2) addresses the possibility of missed detections due to sensor faults and the positive/nuisance false alarms that any real sensor can generate. In this chapter we describe two techniques that can be adopted to deal with those uncertainties. The first technique is based on Event Tree heuristic distance metrics. It allows to generate warnings whenever a threat scenario is detected and it is similar to the ones in the knowledge base repository. The second technique allows to measure in real-time the estimated trustworthiness of event detection based on: a) sensors false alarm rates; b) uncertainties indices associated to correlation operators. We apply those techniques to case-studies of physical security for metro railways.

Keywords: Physical Security Information Management, Dependability, Situation Recognition, False Alarms, Soft Computing, Fuzzy Logic.

1 Introduction

Modern surveillance solutions for infrastructure protection are based on the integration of different sensing subsystems. Each subsystem can include a large number of diverse and/or redundant distributed sensors, which are in charge of detecting abnormal conditions or unwanted events in the monitored environment.

W. Zamojski and J. Sugier (eds.), *Dependability Problems of Complex Information Systems*, Advances in Intelligent Systems and Computing 307, DOI: 10.1007/978-3-319-08964-5_3

The rational exploitation of the available sensing capabilities needs a proper management and processing of both the "modeled" and "captured" information together with the related uncertainty. Therefore, together with PSIM systems there is an increasing need for the appropriate management of parameters characterizing sensor performances (see references [1,5,6]).

Ideally, the sensors should detect only "real" alarms, that represent a true threat. However, many devices generate unnecessary warnings, which can be classified as false alarms or nuisance alarms. False alarms are due to events that should not cause an alarm, while nuisance alarms are generated when a legitimate cause occurs, but alarm activation is not due to a real threat. The same consideration is still valid for a sensing subsystem as a whole, i.e. including sensing devices and specific software for the processing of what they detect (e.g. intelligent video surveillance systems include cameras and video content analytics for the detection of events).

We have addressed the issue of automatic situation recognition by developing a framework for model-based event correlation in infrastructure surveillance. The framework – named DETECT – is able to store in its knowledge base any number of threat scenarios described in the form of Event Trees, and then recognize those scenarios in real-time, providing early warnings to PSIM users [7].

The aim of this chapter is to provide means to improve both effectiveness and efficiency of situation recognition in PSIM. Effectiveness is achieved by enriching the system with enhanced detection capabilities by defining and computing appropriate Event Tree distance metrics based on heuristic approaches. That allows to reduce the number of threats to be modelled and included in the knowledge base (i.e. scenario repository). Efficiency is to be intended at human-machine interaction level, by associating a level of trustworthiness to threat detection in order to allow PSIM operators to be aware of alarm credibility and priority of intervention, and hence react consequently.

More specifically, we can evaluate the impact of the reliability of each sensor/subsystem on the reliability of the whole integrated surveillance system, in terms of POD (Probability of Detection) and FAR (False Alarm Rate) parameters. The first characterizes the effectiveness of a detection system, the second determines its operational viability [3,4]. The need for such an evaluation is especially important when integrated surveillance systems are extended by means of a correlation engine aimed at the automatic threat detection and situation recognition. In fact, in that case, the alarm activation is based on the correlation of different sensors output (and involves also the activation of the related countermeasures).

In order to demonstrate the application of the approach, several threat scenarios impacting physical security of metro railway stations are considered.

The rest of this chapter is structured as follows. Section 2 provides an overview of the related literature on DETECT and on the topics addressed by this chapter, and it introduces the basic concepts of the event description language. Section 3 describes the metrics used to evaluate the distance between event trees and provides several application examples in DETECT. Section 4 describes the on-line computation of detection trustworthiness using sensor performance and reliability data, together with a possible "fuzzy" correlation approach. Finally, Section 5 provides conclusions and hints for future improvements.

2 Background

2.1 Related Works

The first concept of DETECT has been described in [7], where the overall architecture of the framework is presented, including the composite event specification language (EDL, Event Description Language), the modules for the management of detection models and the scenario repository. In [5], an overall system including a middleware for the integration of heterogeneous sensor networks is described and applied to railway surveillance case-studies. Reference [14] discusses the integration of DETECT in the PSIM system developed by Ansaldo STS, namely RailSentry [2], presenting the reference scenario which will be also used in this chapter. The first idea of using scenario similarity analysis has been introduced in reference [15]; however, it only worked with isomorphic trees and therefore its practical utility was rather limited.

A survey of state-of-the-art methods in physical security technologies and advanced surveillance paradigms, including a section on PSIM systems, is provided in [16]. Contemporary remote surveillance systems for public safety are also discussed in [17]. Technology and market-oriented considerations on PSIM can be also found in [18,19].

We could not find any specific applications of real-time trustworthiness evaluation for PSIM like the one we describe in this chapter. However, we report in the following some "static" approaches dealing with multi-sensor dependability evaluation.

In [8] the authors address the issue of providing fault-tolerant solutions for WSN, using event specification languages and voting schemes. A similar issue is addressed in [9], where the discussion focuses on different levels of information/decision fusion on WSN event detection using appropriate classifiers and reaching a consensus among them in order to enhance trustworthiness. Reference [13] describes a method for evaluating the reliability of WSN using the Fault Tree modelling formalism, but the analysis is limited to hardware faults (quantified by the Mean Time Between Failures, MTBF) and homogenous devices (i.e. the WSN motes). Performance evaluation aspects of distributed heterogeneous surveillance systems are instead addressed in [11], which only lists the general issues and some pointers to the related literature. Reference [10] addresses trustworthiness analysis of sensor networks in cyber-physical systems, focusing on the reduction of false alarms by clustering sensors according to their locations and by building appropriate object-alarm graphs. Another general discussion on the importance of the evaluation of performance metrics and human factors in distributed surveillance systems can be found in [12].

2.2 Event Description Language

Threats scenarios are described in DETECT using a specific Event Description Language (EDL) and stored in a Scenario Repository. In this way we are able to store permanently all scenario features in an interoperable format (i.e. XML). A high level architecture of the framework is depicted in Fig. 1.

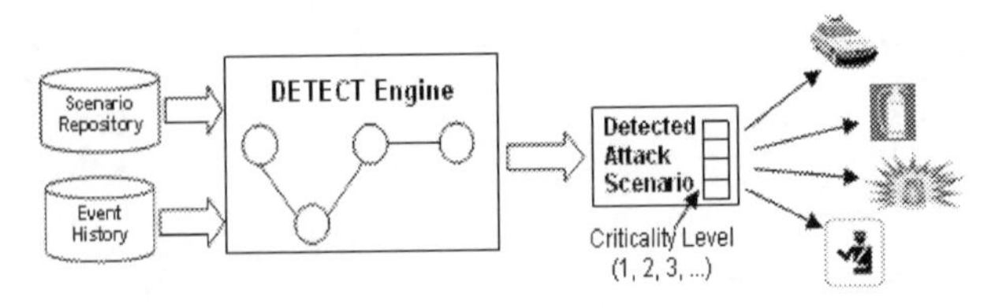

Fig. 1. The DETECT framework

A threat scenario expressed by EDL consists of a set of basic events detected by the sensing devices. An event is a happening that occurs at some locations and at some points in time. In this context, events are related to sensor data (e.g. temperature higher than a threshold, motion detected by an intelligent camera, intrusion detected by a volumetric sensor). Events are classified as *primitive events* and *composite events*.

A primitive event is a condition on a specific sensor which is associated to some parameters (i.e. event identifier, time of occurrence, etc...). A composite event is a combination of primitive events by means of proper logical and temporal operators.

Each event is denoted by an *event expression*, whose complexity grows with the number of involved events. Given the expressions $E_1, E_2, ..., E_n$, every application on them through any operator is still an expression. Event expressions are represented by *Event Trees*, where primitive events are at the leaves and internal nodes represent EDL operators.

DETECT is able to support the composition of complex events in EDL through a *Scenario GUI* (Graphical User Interface), used to draw threat scenarios by means of a user-friendly interface.

Furthermore, in the operational phase, a model manager macro-module has the responsibility of performing queries on the Event History database for the real-time feeding of detection models corresponding to threat scenarios, according to predetermined policies. Those policies, namely *parameter contexts*, are used to set a specific consumption mode of the occurrences of the events collected in the database.

The EDL is based on the Snoop event algebra [20], considering the following operators: OR, AND, ANY, SEQ. As an example, Fig. 2 shows a simple event tree representing the scenario (E_1 AND E_2) OR E_3. In this example scenario, for the sake of simplicity, the association between the primitive event and the sensor, which detected it, is not made explicit.

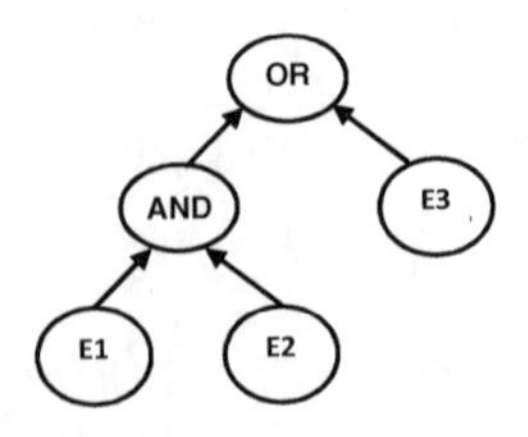

Fig. 2. A simple event tree

The semantics of the Snoop operators are as follows:

- **OR.** Disjunction of two events E_1 and E_2, denoted $(E_1\ OR\ E_2)$. It occurs when at least one of its components occurs.
- **AND.** Conjunction of two events E_1 and E_2, denoted $(E_1\ AND\ E_2)$. It occurs when both events occur (the temporal sequence is ignored).
- **ANY.** A composite event, denoted $ANY\ (m, E_1, E_2, ..., E_n)$, where $m \leq n$. It occurs when m out of n distinct events specified in the expression occur (the temporal sequence is ignored).
- **SEQ.** Sequence of two events E_1 and E_2, denoted $(E_1\ SEQ\ E_2)$. It occurs when E_2 occurs provided that E_1 has already occurred. This means that the time of occurrence of E_1 has to be less than the time of occurrence of E_2.

Furthermore, *temporal constraints* can be specified on operators, to restrict the time validity of logic correlations.

In order to take into account appropriate event consumption modes and to set how the occurrences of primitive events are processed, four parameter contexts are defined. Given the concepts of *initiator* (the first constituent event whose occurrence starts the composite event detection) and *terminator* (the constituent event that is responsible for terminating the composite event detection), the four different contexts are described as follows.

- *Recent*: only the most recent occurrence of the initiator is considered.
- *Chronicle:* the (initiator, terminator) pair is unique. The oldest initiator is paired with the oldest terminator.
- *Continuous:* each initiator starts the detection of the event.
- *Cumulative*: all occurrences of primitive events are accumulated until the composite event is detected.

The effect of the operators is then conditioned by the specific context in which they are placed.

When a composite event is recognized, the output of DETECT consists of:

- the identifier(s) of the detected/suspected scenario(s)[1];
- the temporal value related to the occurrence of the composite event (corresponding to the event occurrence time of the last component primitive event, given by the sensor timestamp);

[1] The difference between detected and suspected scenario depends on the partial or total matching between the real-time event tree and the stored threat pattern.

- an alarm level (optional), associated to scenario evolution (used as a progress indicator and set by the user at design time);
- other information depending on the detection model (e.g. 'likelihood' or 'distance', in case of heuristic detection).

3 Heuristic Distance Metrics for Event Trees

3.1 Definition of Distance Metrics

The following attributes can be associated to Event Trees (positive integer numbers):

1. *TN*: total number of nodes
2. *TD*: tree depth, that is the number of levels from leaves to the top node
3. *TW*: tree width, that is the maximum number of operators at the same level
4. *SL*: set of leaf nodes
5. *SO*: set of operator nodes

Though other attributes (e.g. number of arcs) could be associated to event trees, the ones listed above seem to picture a comprehensive yet not redundant set of characteristics. While a theoretical demonstration could be possible, such a statement has been validated experimentally. For instance, the number of arcs in all the significant scenarios included in the repository was always dependant on the number of nodes.

In order to obtain an easy to compute metric, the distance between two event trees is obtained as the sum of the differences between homologous attributes. In other words, the distance D among event trees A and B is obtained as follows:

$$D = |TN_A - TN_B| + |TD_A - TD_B| + |TW_A - TW_B| + DSL_{AB} + DSO_{AB}$$

(+ 1 if parameter contexts are different)

The quantities DSL and DSO are computed as set differences (*card* competes the cardinality of the set):

$$DSL_{AB} = card(SL_A \cup SL_B) - card(SL_A \cap SL_B)$$

$$DSO_{AB} = card(SO_A \cup SO_B) - card(SO_A \cap SO_B)$$

It is quite obvious that such a heuristic distance metric can be applied to any couple of event trees, regardless of possible isomorphisms[2].

[2] Two trees are isomorphic when they are identical in graph structure (they could differ in node attributes).

3.2 Implementation in DETECT

An appropriate algorithm has been implemented in DETECT in order to properly compute tree attributes. In the following, the notation of each leaf node (i.e. Ex-Sx, where x is a positive integer) includes indications regarding both the typology of the primitive event (i.e. Ex) and the sensor responsible for its detection (i.e. Sx), according to a pre-defined encoding.

Starting from the root node, the whole tree is scanned and each node is saved in a table where each row represents a tree level (see Fig. 3). For each node, the name of the father node is saved as well as the list including the names of the son nodes. In the end of the scan, all the information relevant for tree attributes computation will be available. Hence the formula to obtain the distance between any couple of trees can be easily computed. Off-line distance calculation is very useful when inserting a new event tree in the Scenario Repository. In fact, when a human operator finishes building the event tree and saves it in the repository, he/she can see all the distances (possibly only the ones lower than a certain threshold) with all the other event trees in the repository. Therefore, if another tree exists whose distance from the new one is very low, then it is possible the two trees represent the same threat (or similar threats) and therefore could be somehow merged to reduce multiple warnings and improve usability as well.

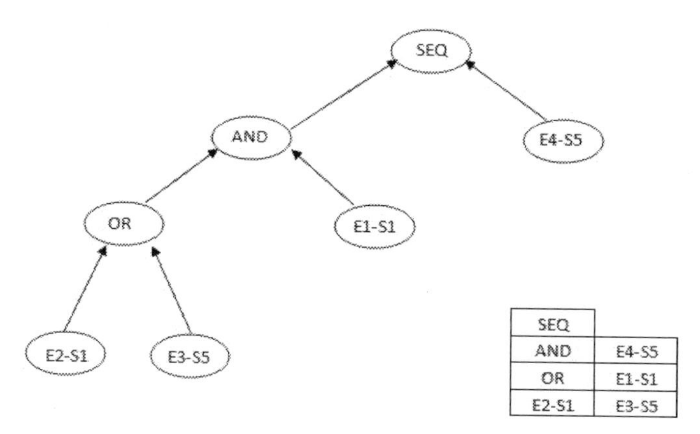

Fig. 3. An example table obtained from an event tree

For on-line calculation, the distance needs to be computed bottom-up starting from subtree attributes, which will be associated at run-time to each node (see Fig. 4). Due to the working logics of DETECT, some limitations hold for the run-time computation of tree attributes (e.g. the *TD* metric cannot be computed at run-time). More specifically, since operator nodes can be considered as the roots of the subtrees below them, it is possible to associate to operator nodes the attributes of the subtrees below them. Hence, moving from the leaves to the root and exploiting the already computed attributes, each operator node will be associated to updated attributes representing all the tree structure below it. Therefore, the root node will include the overall attributes of the whole tree. When a subtree is detected and its alarm level in

DETECT is greater than 0, its attributes are compared with the ones of all the other event trees in the Scenario Repository. If the distance D with another threat scenario T is lower than a configurable threshold D_T, then a warning is generated and shown to the PSIM human operator, in order to warn him/her about the risk that threat T is occurring. It is obviously possible to associate different warning levels to different distances (the lower the distance metric, the highest the warning level); however, in practical applications it is important to keep the system simple to understand to operators. Therefore, we have decided to use a single threshold and a single warning level.

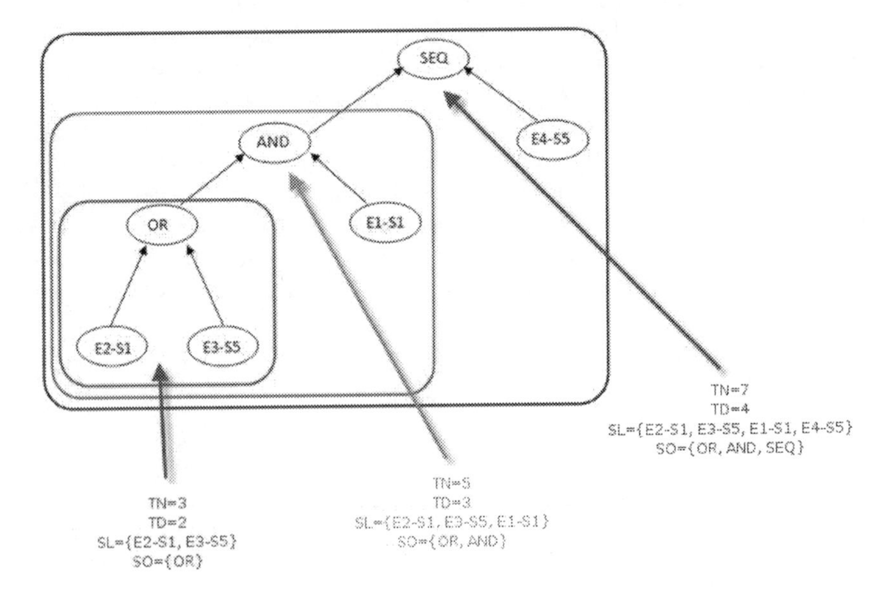

Fig. 4. Example of on-line subtree attributes computation

3.3 Distance Metrics Examples

In this section we report some examples of evaluation of attributes and distance metrics for reference threat scenarios. The first scenario we consider is the Chemical Attack (scenario A) by means of a CWA (Chemical Warfare Agent), which we have already described in reference [14], whose event tree is depicted in Fig. 5 together with a table including its attributes.

Events in the scenario are described as follows using the notation "sensor description (sensor ID) :: event description (event ID)":

Intelligent Camera (S1) :: Fall of person (E1)
Intelligent Camera (S1) :: Abnormal running (E2)
Intelligent Camera (S2) :: Fall of person (E1)
Intelligent Camera (S2) :: Abnormal running (E2)

Audio sensor (S3) :: Scream (E3)
IMS/SAW detector (S4) :: CWA detection (E4)
IR detector (S5) :: CWA detection (E4)

The same scenario could be represented in other way using the model of Fig. 6 (scenario B), featuring slightly different attributes.

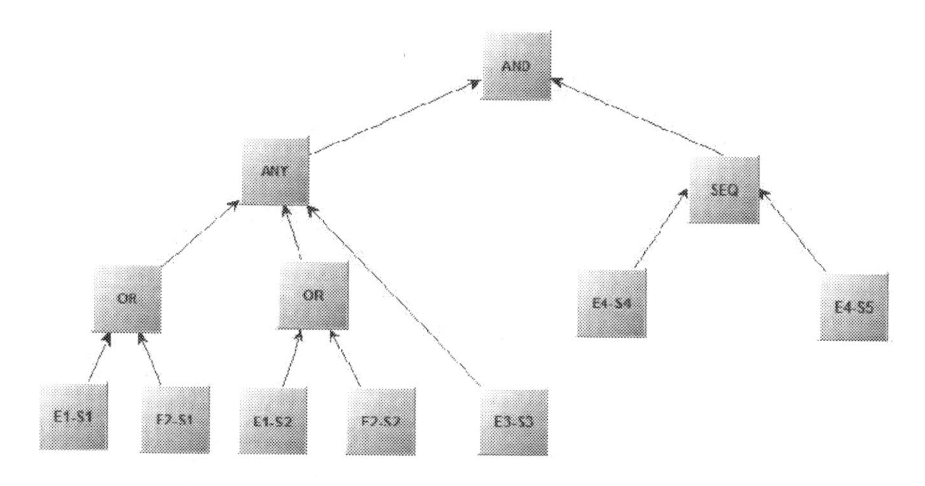

SCENARIO A

TN	12	
TD	3	
TW	2	
SL	E1-S1, E2-S1, E1-S2, E2-S2, E3-S3, E4-S4, E4-S5	cardinality=7
SO	AND, ANY, SEQ, OR	cardinality=4

Fig. 5. Event tree attributes for the Chemical Attack scenario

The two scenarios of Fig. 5 and Fig. 6 feature the same primitive events (i.e. the trees have the same leaves) and therefore the SL distance is 0. Instead, the sets of operators differ by 1. Overall, the distance is given by:

$$D = |12-10| + |3-3| + |2-1| + 0 + 1 = 4$$

Now, let us consider a scenario of pickpocketing/aggression (scenario C), which could partially overlap with the previous one regarding people behaviour, since it features the composite event represented by the ANY operator included in scenario B. Furthermore it is similar to the corresponding ANY in scenario A. Please refer to Fig. 7, where E5-S6 represents an alarm coming from the emergency call point.

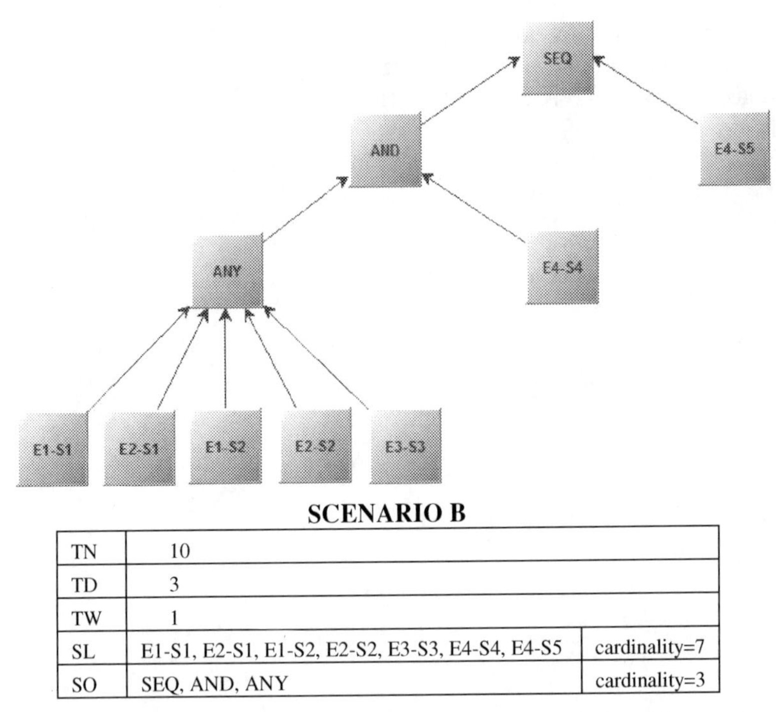

TN	10	
TD	3	
TW	1	
SL	E1-S1, E2-S1, E1-S2, E2-S2, E3-S3, E4-S4, E4-S5	cardinality=7
SO	SEQ, AND, ANY	cardinality=3

Fig. 6. Event tree attributes for another version of the Chemical Attack scenario

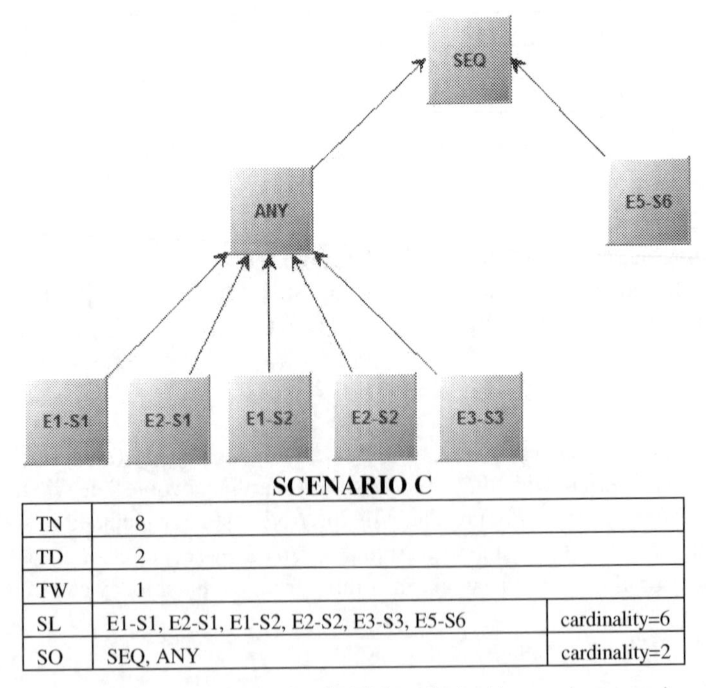

TN	8	
TD	2	
TW	1	
SL	E1-S1, E2-S1, E1-S2, E2-S2, E3-S3, E5-S6	cardinality=6
SO	SEQ, ANY	cardinality=2

Fig. 7. Event tree attributes for the Pickpocketing/Aggression scenario

Table 1. Differences among attributes of scenarios A, B and C

	A-B	**A-C**	**B-C**
ΔTN	2	4	2
ΔSL	0	3	3
ΔTD	0	1	1
ΔSO	1	2	1
ΔTW	1	1	0
D	4	11	7

An overview of distances among attributes of event trees representing scenarios A, B and C is reported in Tab. 1.

As an example, in off-line operation, when inserting scenario B after A and C, the human operator sees the distances with scenarios A and C. In this case, he/she will be aware of the similarity with scenario A, since the distance is low (e.g. D_T could be set to 5) and could decide to keep only the original version (i.e. scenario A) since the variation would be automatically detected by the on-line heuristic engine based on distance.

In on-line operation, let us assume the ANY event of scenario A is detected. The expected behaviour will be as follows.

1. DETECT computes the attributes associated to the ANY composite event subtree (see below).

TN	8	
TD	2	
SL	E1-S1, E2-S1, E1-S2, E2-S2, E3-S3	cardinality=5
SO	ANY, OR	cardinality=2

2. DETECT computes the distances with all the (enabled and full) event trees in the Scenario Repository (see D row below).

	ANY-A	**ANY-B**	**ANY-C**
ΔTN	4	2	0
ΔSL	2	2	1
ΔTD	1	1	0
ΔSO	2	3	1
D	9	8	2

The computed distances correctly represent the recognised situation that, though formally belonging to scenario A, in absence of chemical warfare agent detection, is more similar to a situation of aggression/pickpocketing.

Given the possibility to get additional, but still appropriate warnings on possible forthcoming threats, the on-line operation is strategic to enrich the detection capabilities of the existing deterministic correlation engine. In particular, the described recognition technique addresses the imperfect threat modeling, due to human faults, as well as the possible missed detections, due to sensor faults.

4 Measuring Detection Trustworthiness in Real-Time

In this section we introduce a further additional feature to take into account the uncertainty due to sensor false alarms. In particular we describe how to exploit the parameters describing the detection performance[3] of the sensors involved in physical security situation recognition, in order to evaluate the trustworthiness of the inferred alarms. Addressing such an issue is very important in physical security management systems, where the alarms are sent to a control center and the triggering of countermeasures can be fully automatic (independent from human intervention) or partially automatic (based on human discrimination).

In order to associate a reliability level to event detection, it is possible to use a real-time fuzzy correlation of sensor outputs using a Bayesian Network (BN). Such a probabilistic modeling formalism enables a fuzzy logic through the use of "noisy" logic gates, whenever the output is not deterministic, but associated with a certain probability [21].

Formally, let us define a *detector* as a sensor or a sensing subsystem which in relation to a certain event can provide two outputs:

- TRUE – if the event has been detected;
- FALSE – if not.

Each detector can be associated to the following parameters:

- POD = P(event detected | event occurred);
- FAR = P(event detected | event not occurred).

An analysis based on the POD of detectors can be used to compute the probability in threat recognition, while we build the related detection models. Therefore it is convenient at design-time, since the results can provide a guide to draw appropriate event trees and to support the choice and dislocation of detectors, with respect to the specific threats to be addressed. The main end is to reach a certain target in the probability of recognition a particular threat, before using its detection model at real time. Such an analysis is objective of another work and it is not described in this chapter. Let us to address a FAR based real-time analysis in the following.

Assuming the use of AND and OR logical operators in order to correlate the outputs of detectors, we can perform an analysis based on their FAR parameters and aimed at the calculation of alarms reliability in real-time. A synthetic indicator of such an evaluation can then be shown to PSIM operators together with alarms.

[3] In this section we refer to detection performance, reliability and trustworthiness by meaning the same concept related to false alarm generation (i.e. false positive).

Table 2. Probabilistic parameters of two possible sensors

Detector ID	Event ID	FAR
S1	E1	0.15
S2	E2	0.10

In the following example, we assume using detectors whose FAR is described in Tab. 2. With reference to the AND operator, we can model the alarm reliability through a simple Bayesian Network (see Fig. 8).

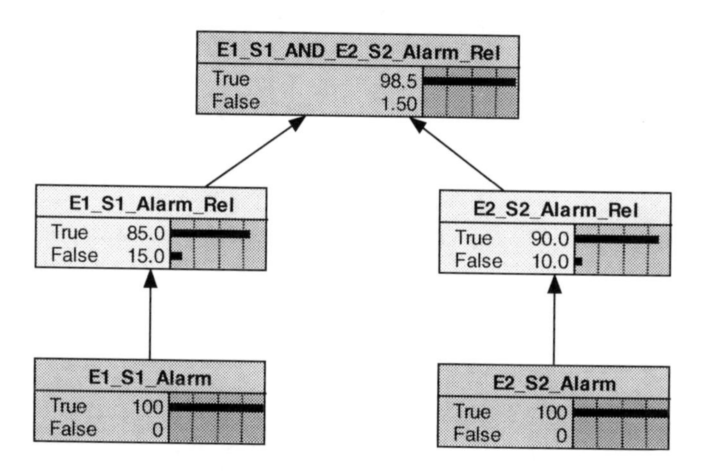

Fig. 8. Example of BN modelling an AND logical operator

The leaf nodes represent the occurrence of the alarms associated to the events E*x* detected by S*x*. The reliability of each alarm (E*x*_S*x*_Alarm_Rel) is calculated using the FAR parameter of the related detector by using the following formula:

P(E*x*_S*x* is TRUE | E*x*_S*x*_Alarm has been generated) =
= P(E*x*_S*x*_Alarm is not FALSE) = 1 – P(E*x*_S*x*_Alarm is FALSE) = 1 – FAR_{Ex_Sx}

The alarms reliability reported in the BN are represented in percentages. The CPT (Conditional Probability Table) of the AND node is reported in Tab. 3.

Table 3. CPT of the AND node

E1_S1_Alarm_Rel	E2_S2_Alarm_Rel	E1_S1_AND_E2_S2_Alarm_Rel
True	True	True
True	False	True
False	True	True
False	False	False

The following hypothesis holds: the alarm associated to the AND event is not considered reliable only if both the alarms associated to E1_S1 and E2_S2 events are not considered reliable. For example, it means that when S1 detects E1 correctly and S2 generates a false alarm in E2 detection, then the related AND event – which is triggered anyway – is classified as TRUE. However, by modifying the CPT properly we can consider a more conservative hypothesis: the alarm associated to the AND event is considered TRUE only if both the alarms associated to E1_S1 and E2_S2 events are considered reliable. In the first case (shown in Fig. 8) we have an AND alarm reliability of 98.5%, in the second one, we have a lower value (76.5%).

The real-time calculation of an OR alarm reliability is quite simple, since OR alarm activation is concomitant with the single Ex_Sx alarm generated first (i.e. 85% or 90% depending on the case).

The effectiveness of the approach increases significantly when we consider complex Event Trees. In those scenarios, when basic events are detected by sensors, they feed detection models according to the scenario evolution. Thus, step by step, the BN related to each occurred subtree can be executed in real-time in order to get also the alarm reliability related to the inferred composite event. In that case, the approach is easy to apply also to the other operators. In fact, in real-time analysis, the SEQ (sequence) operator can be treated as an AND, while ANY(k,n) operator (that is equivalent to the "k out of n" scheme) can be treated as an n-ary AND.

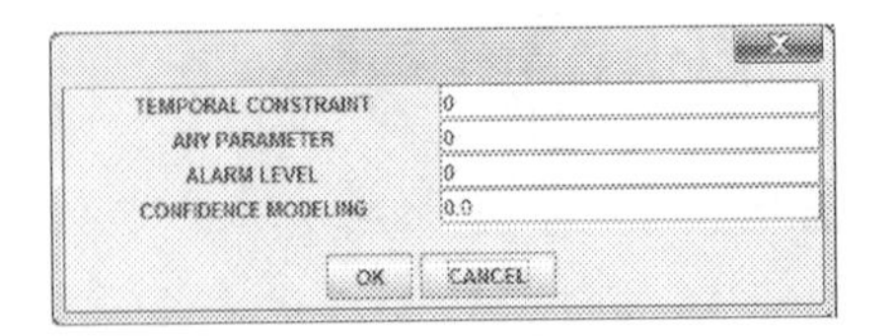

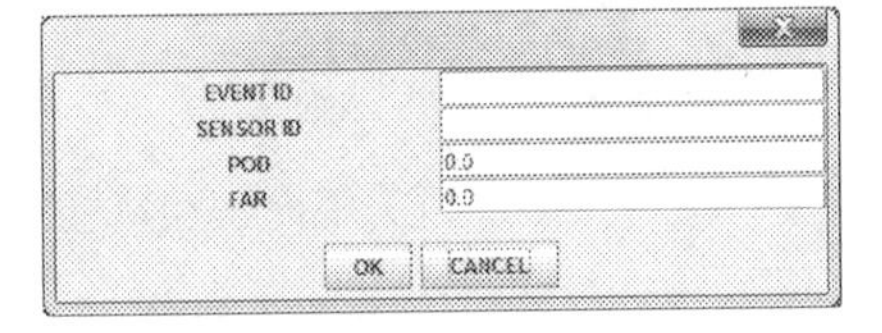

Fig. 9. DETECT entry windows for operator and basic event parameters

Finally, we can take into account also the uncertainty of the detection models used to recognize threat scenarios. More in detail, in order to consider a possible mismatching between a real threat scenario and its model, for each logical operator we can set also a confidence index, which weighs the trustworthiness of the operator. In other words, if we set the index to a probability value p in the range from 0 to 1 (1 is the default value representing no uncertainty), then the occurrence of the logical condition represented by the operator is True with a probability p weighted with the computed alarm reliability. All the input parameters we have described in this chapter can be entered in DETECT framework by means of proper windows of its GUI (Graphical User Interface) shown in Fig. 9. The whole logic based on BNs is therefore completely transparent to the user and it is fully integrated with the one based on Event Trees.

A practical application of the approach is described as follows. Let us consider the chemical attack scenario already addressed in the previous section (scenario A), which describes the drop of a CWA in a metro railway platform, represented by the event tree in Fig. 5. The scenario is built considering two intelligent cameras positioned at platform end walls, a microphone between them, two standoff detectors for CWAs positioned on the platform and on the escalator or concourse level. Let us assume to characterize the involved detectors with the FAR parameters reported in Tab. 4.

Table 4. FAR parameters of detectors used in chemical attack scenario

Detector ID	Detector Description	Event ID	Event Description	FAR
S1	Intelligent Camera	E1	Fall of person	0.25
		E2	Abnormal running	0.20
S2	Intelligent Camera	E1	Fall of person	0.25
		E2	Abnormal running	0.20
S3	Audio Sensor	E3	Scream	0.15
S4	IMS/SAW detector	E4	CWA detection	0.30
S5	IR detector	E4	CWA detection	0.27

Please note that single events detected by intelligent cameras do not represent necessarily a threat situation. In the approach we are describing, a low alarm level (e.g. to 1) can be associated to the OR operators. When 2 out of 3 distinct events detected by intelligent cameras and/or microphone occur, the monitored situation is considered abnormal. So the alarm level of the ANY operator is set to 2. The use of the sequence operator is due to the different locations of the CWA detectors: IMS/SAW detector (combining Ion Mobility Spectroscopy and Surface Acoustic Wave technologies) at platform level, and IR detector (based on Infrared Radiation) at escalator or concourse level, in such a way to detect correctly the spread of CWA and avoid possible false alarm conditions. The alarm level of the SEQ operator is set to 3. Finally, the detection of the whole threat scenario is associated to the AND occurrence. Its alarm level is set to 4. The use of many alarm levels is strategic to trigger countermeasures properly. Further details on the modeling of the chemical attack scenario are described in [14].

A possible set of basic event occurrences corresponding to a real CWA attack is listed in Tab. 5, which includes chronological aspects like the ones used in real PSIM log-files.

Table 5. A possible basic events chronology related to the CWA attack

Date	Time	Event ID	Detector ID	Occurrence Nr
01/04/2012	09:11:11	E4	S4	1
01/04/2012	09:14:18	E1	S2	2
01/04/2012	09:15:51	E3	S3	3
01/04/2012	09:16:00	E2	S2	4
01/04/2012	09:17:07	E4	S5	5

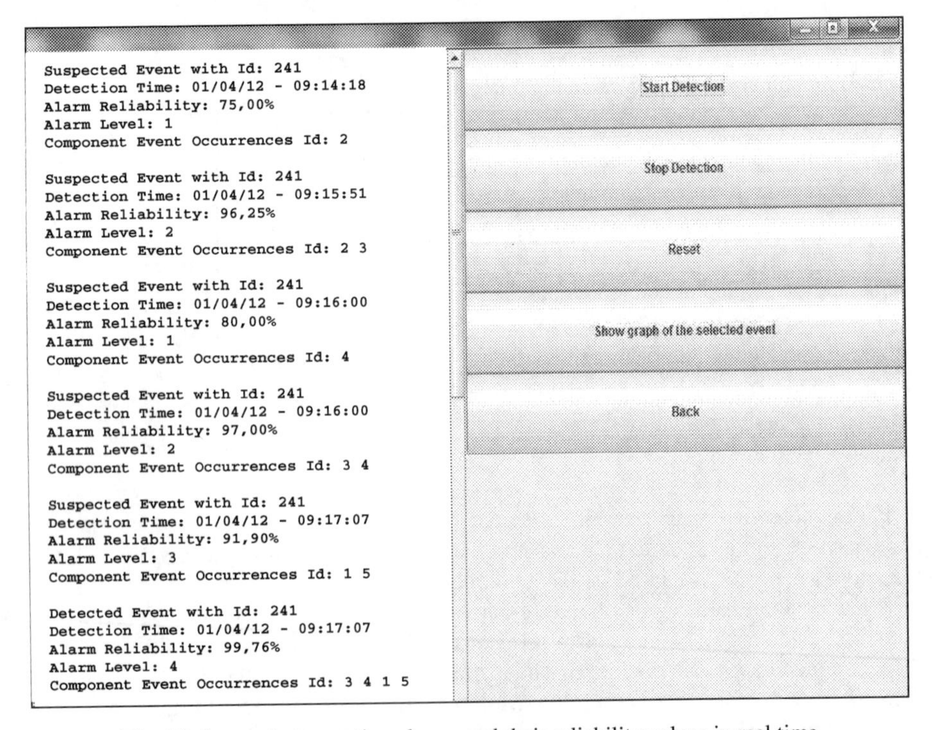

Fig. 10. Screenshot reporting alarms and their reliability values in real time

When using DETECT to model the threat scenario (whose ID is assigned - for example - 241), with the Event Tree of Fig. 5 and the parameters of Tab. 4, the output is reported in the screenshot in Fig. 10: for each detected event, also the reliability level is reported, which is calculated in real-time using the BN approach. In the described example we have considered no uncertainty coming from the detection model (the confidence index of each operator used to build the event tree is set to its default value, i.e. 1)

The real-time execution of the BN models also enables the possibility of using 'dynamic' FAR parameters, continuously updated using the feedback of the PSIM operators in terms of confirmation of the alarms detected in the real on-the-field operation. In other words, for each event detected by a sensor, the statistical analysis of the ratio ('false positive alarms' / 'total number of alarms'), can lead to a proper update of FAR parameters and therefore to more reliable estimations with respect to the 'static' ones.

5 Conclusions and Future Work

In this chapter we have provided some means to address the problem of uncertainty management in threat detection with PSIM systems. The analysis was based on a reference framework that has been developed to correlate alarms coming from heterogeneous detectors, namely DETECT. DETECT uses Event Trees as its main modeling formalism and therefore as the base for its correlation engine. The results described in this chapter are limited to threat scenarios specified with Event Trees. Event Trees has the advantage of being a simple modeling formalism for physical security threat modeling, but also has some limitations. The main limitations, related to the difficulty in managing uncertainty in model structure and parameters, have been addressed in this chapter in order to be able to use the framework in dependable surveillance applications. To that aim, we have developed a set of threat scenarios relevant for metro railway contexts, some of which have been briefly addressed in this chapter for the case-study applications.

Though the main problems have been solved, we envisage some further developments mainly aimed at the automatic 'learning' of uncertainty parameters using the feedback provided by PSIM operators in the operational stage. That would allow a continuous update of the knowledge base to fine-tune performance and dependability in real-world surveillance applications. Also, other dependability related aspects of complex situation recognition in surveillance systems have been recently addressed in reference [22], which focuses on probabilistic models for the static evaluation of threat detection trustworthiness on reference scenarios, also allowing quantitative analysis of model sensitivity to sensor faults.

References

1. Garcia, M.L.: The Design and Evaluation of Physical Protection Systems. Butterworth-Heinemann (2001)
2. Bocchetti, G., Flammini, F., Pragliola, C., Pappalardo, A.: Dependable integrated surveillance systems for the physical security of metro railways. In: IEEE Procs. of the Third ACM/IEEE International Conference on Distributed Smart Cameras, pp. 1–7 (2009)
3. Zhu, Z., Huang, T.S.: Multimodal Surveillance: Sensors, Algorithms and Systems. Artech House Publisher (2007)
4. Wickens, C., Dixon, S.: The benefits of imperfect diagnostic automation: a synthesis of the literature. Theoretical Issues in Ergonomics Science 8(3), 201–212 (2007)
5. Flammini, F., Gaglione, A., Mazzocca, N., Moscato, V., Pragliola, C.: Wireless Sensor Data Fusion for Critical Infrastructure Security. In: Corchado, E., Zunino, R., Gastaldo, P., Herrero, Á. (eds.) CISIS 2008. ASC, vol. 53, pp. 92–99. Springer, Heidelberg (2009)

6. Flammini, F., Gaglione, A., Ottello, F., Pappalardo, A., Pragliola, C., Tedesco, A.: Towards Wireless Sensor Networks for Railway Infrastructure Monitoring. In: Proc. ESARS 2010, Bologna, Italy, pp. 1–6 (2010)
7. Flammini, F., Gaglione, A., Mazzocca, N., Pragliola, C.: DETECT: a novel framework for the detection of attacks to critical infrastructures. In: Martorell, et al. (eds.) Procs. of ESREL 2008, pp. 105–112 (2008)
8. Ortmann, S., Langendoerfer, P.: Enhancing reliability of sensor networks by fine tuning their event observation behaviour. In: Proc. 2008 International Symposium on a World of Wireless, Mobile and Multimedia Networks (WOWMOM 2008), pp. 1–6. IEEE Computer Society, Washington, DC (2008)
9. Bahrepour, M., Meratnia, N., Havinga, P.J.M.: Sensor Fusion-based Event Detection in Wireless Sensor Networks. In: 6th Annual International Conference on Mobile and Ubiquitous Systems: Networking and Services, MobiQuitous 2009, Toronto, Canada (2009)
10. Tang, L.-A., Yu, X., Kim, S., Han, J., Hung, C.-C., Peng, W.-C.: Tru-Alarm: Trustworthiness Analysis of Sensor Networks in Cyber-Physical Systems. In: Proceedings of the 2010 IEEE International Conference on Data Mining (ICDM), IEEE Computer Society, Washington (2010)
11. Legg, J.A.: Distributed Multisensor Fusion System Specification and Evaluation Issues. Defence Science and Technology Organisation, Edinburgh, South Australia 5111, Australia (2005)
12. Karimaa, A.: Efficient Video Surveillance: Performance Evaluation in Distributed Video Surveillance Systems. In: Surveillance, V., Lin, W. (eds.). InTech (2011), http://www.intechopen.com/books/video-surveillance/efficient-video-surveillance-performance-evaluation-in-distributed-video-surveillance-systems
13. Silva, I., Guedes, L.A., Portugal, P., Vasques, F.: Reliability and Availability Evaluation of Wireless Sensor Networks for Industrial Applications. Sensors 12(1), 806–838 (2012)
14. Flammini, F., Mazzocca, N., Pappalardo, A., Pragliola, C., Vittorini, V.: Augmenting surveillance system capabilities by exploiting event correlation and distributed attack detection. In: Tjoa, A.M., Quirchmayr, G., You, I., Xu, L. (eds.) ARES 2011. LNCS, vol. 6908, pp. 191–204. Springer, Heidelberg (2011)
15. Flammini, F., Pappalardo, A., Pragliola, C., Vittorini, V.: A robust approach for on-line and off-line threat detection based on event tree similarity analysis. In: Proc. Workshop on Multimedia Systems for Surveillance (MMSS) in Conjunction with 8th IEEE International Conference on Advanced Video and Signal-Based Surveillance, pp. 414–419 (2011)
16. Flammini, F., Pappalardo, A., Vittorini, V.: Challenges and emerging paradigms for augmented surveillance. In: Effective Surveillance for Homeland Security: Balancing Technology and Social Issues, pp. 169–198. Taylor & Francis/CRC Press (2013)
17. Räty, T.D.: Survey on contemporary remote surveillance systems for public safety. IEEE Trans. Sys. Man Cyber Part C 5(40), 493–515 (2010)
18. Hunt, S.: Physical security information management (PSIM): The basics, http://www.csoonline.com/article/622321/physical-security-information-management-psim-the-basics
19. Frost, Sullivan: Analysis of the Worldwide Physical Security Information Management Market (2012), http://www.cnlsoftware.com/media/reports/Analysis_Worldwide_Physical_Security_Information_Management_Market.pdf

20. Chakravarthy, S., Mishra, D.: Snoop, An expressive event specification language for active databases. Data Knowl. Eng. 14(1), 1–26 (1994)
21. Ben Mrad, A., Maalej, M.A., Delcroix, V., Piechowiak, S., Abid, M.: Fuzzy Evidence in Bayesian Network. In: Proc. Intl Conf. on Soft Computing and Pattern Recognition, pp. 486–491 (2011)
22. Flammini, F., Marrone, S., Mazzocca, N., Pappalardo, A., Pragliola, C., Vittorini, V.: Trustworthiness Evaluation of Multi-sensor Situation Recognition in Transit Surveillance Scenarios. In: Cuzzocrea, A., Kittl, C., Simos, D.E., Weippl, E., Xu, L. (eds.) CD-ARES Workshops 2013. LNCS, vol. 8128, pp. 442–456. Springer, Heidelberg (2013)

Testing-as-a-Service for Mobile Applications: State-of-the-Art Survey

Oleksii Starov[1], Sergiy Vilkomir[2], Anatoliy Gorbenko[3], and Vyacheslav Kharchenko[3]

[1] Computer Science Department,
State University of New York at Stony Brook, USA
ostarov@cs.stonybrook.edu
[2] Department of Computer Science,
East Carolina University, USA
vilkomirs@ecu.edu
[3] Department of Computer Systems and Networks (503)
National Aerospace University, Kharkiv, Ukraine
A.Gorbenko@csn.khai.edu,
V.Kharchenko@khai.edu

Abstract. The paper provides an introduction to the main challenges in mobile applications testing. In the paper we investigate the state-of-the-art mobile testing technologies and overview related research works in the area. We discuss general questions of cloud testing and examine a set of existing cloud services and *testing-as-a-service* resources facilitating testing of mobile applications and covering a large range of the specific mobile testing features.

Keywords: Mobile application, software testing, cloud services.

1 Introduction

Mobile development is characterized by a variety of applications with different quality requirements. Online application stores, like the Apple App Store and Google Play, offer thousands of market-oriented apps—mobile games, utilities, navigators, social networks, and clients for web resources. At the same time, the interest in critical mobile applications is growing. For instance, online banking has evolved into mobile banking, mobile social alerts are widely used to report accidents or warn about hurricanes [1], and special apps exist to monitor traffic [2] and help cardiac patients [3]. Augmented reality apps are used for complex navigation and involve a variety of sensors. A new trend is to use smartphones as components for mobile cyber-physical systems because the powerful hardware has a variety of sensors. Mobile applications are even being considered to support processes at such critical facilities as nuclear power plants [4]. These trends require high levels of reliability and quality for mobile software systems. They affect testing, in particular, and the whole mobile development process in general. Too often, the mobile development process ends with the submission of a social application to an online store. The aim is to gain a wider audience of users in a shorter

W. Zamojski and J. Sugier (eds.), *Dependability Problems of Complex Information Systems*,
Advances in Intelligent Systems and Computing 307, DOI: 10.1007/978-3-319-08964-5_4

time, but this does not guarantee the quality of the product and non-critical bugs are usually accepted. Some surveys have confirmed that mobile developers usually deal with small apps and do not adhere to a formal development process [5]. In contrast, a totally different approach is required for critical or business-critical mobile applications, including mobile clients for trustworthy enterprise systems and solutions; for example, Facebook's iOS app is crucial for maintaining the company's profile and reputation and thus was rebuilt to overcome the poor quality of the first version.

To guarantee these mobile applications' reliability and security, sufficient testing is required on a variety of heterogeneous devices as well as on different OS. Android development is the most representative example of how different applications should function amid a plethora of hardware-software combinations [6]. Adequately testing all of these platforms is too expensive—perhaps impossible—especially for small resource-constrained mobile development companies.

Mobile development has a set of distinctive challenges and features. Mobile application testing has some similarities to website testing as both involve validation in many environments (smartphones and browsers, respectively). The general requirements for both types of testing are similar: applications should function correctly, efficiently, and be reliable and secure in all environments. However, mobile testing presents new activities and requires more effort because it includes web applications that work within mobile browsers or hybrid variants wrapped in native code [5]. This testing also involves a large number of possible combinations of mobile devices and OS. Finally, mobile testing involves the use of actual hardware and so testers need additional knowledge and skills such as build installation and crash-log retrieving. Advanced mobile software processes typically work according to the Agile-based methodology [7] and include usage of build distribution services to assist in testing, analytical services for maintenance during production, and services to obtain a wider range of mobile devices for testing. These services create a large set of *testing-as-a-service* (TaaS) resources, or supporting web-applications, that use cloud benefits to facilitate the testing of mobile applications and cover a large range of the specific mobile testing needs. These cloud solutions make mobile testers more effective because they provide complex infrastructure and/or services that are not feasible within small developer companies. The dominant type of such cloud services is a "device cloud," i.e., a service that provides hosting of remote mobile devices and running of tests in the cloud. Existing commercial variants of such platforms became an inspiration for the current study.

The rest of the paper is organized as follows. In the second and third sections we discuss cloud-based testing services and features of mobile applications testing. In Section 4 we systematize existing cloud-based services for mobile application testing including device clouds, application lifecycle management services and discuss techniques for test automation. Additional standalone tools for mobile application testing are described in the Section 5. Finally, the 6th section provides an overview of the modern research studies in mobile testing making emphasis on combinatorial testing techniques in Section 7.

2 Cloud-Based Testing

Many research papers have stated that testing extensively migrates to the cloud nowadays [8–12]. Reviews and classifications of testing cloud services include solutions for web systems and mobile development [13, 14]. Cloud benefits are used not only to support performance, load, or reliability testing of websites, but also to assist with providing required hardware resources (i.e., remote smartphones) for different needs for mobile testing. Cloud-based mobile testing is a young but very topical issue [15].

The database at the Cyber Security and Information Systems Information Analysis Center provides a large list of cloud testing references [16]. Technical and research issues about testing over the cloud are analyzed in [17] and [18] respectively.

In this work we use the term “cloud service” as the most general understanding of cloud computing, i.e., cloud service is a software tool or hardware resource that is delivered over the Internet. The definition means that we also take into account such web resources as build distribution solutions and online issue tracking systems. The term “device cloud” (i.e., mobile device cloud or cloud of devices) will also be used, pointing to both the cloud service’s nature and the many geographically dispersed devices.

Many specialized studies exist regarding the general architecture and construction of cloud and distributed systems [19, 20], including providing service through application programming interfaces (APIs). Technical issues for the tests on the cloud are discussed in [18], including Hadoop usage for test distribution. Device clouds (services that provide hosting of smartphones and run tests on multiple remote real devices) require special algorithms for effective test distribution to make overall test execution time as minimal as possible.

3 Mobile Application Testing

Mobile development has a set of distinctive features and the following specific challenges can be mentioned [5]: support of many hardware and software platforms, correct work with a variety of sensors, interconnections with other applications, high requirements for users’ experiences and the quality of the user interface, and the existence of web mobile and hybrid applications that incorporate all of these challenges to web development.

Mobile applications are popular among startups and approaches for quick prototyping to evaluate the concept of an application are now in high demand. All of these features contribute to the complexity and specifics of mobile testing [6, 21]. As for mobile testing in this work, I mean comprehensive testing of a mobile system that includes the testing of mobile apps as well as mobile operation systems (OS) and the related hardware. Different investigations have pointed to the required mobility of the apps in terms of their ability to function in different environments and configurations as the root challenge of testing [21].

uTest published The Essential Guide to Mobile App Testing [6], a book that comprehensively and coherently describes challenges and techniques in mobile application testing. A lot of research exists about automation and facilitation of the testing process, including leveraging of cloud abilities [10, 22–26]. Companies that provide cloud services for mobile testing (cloud of devices) usually assist their customers with a set of guides [27, 28].

Examples of testing matrixes to cover all smartphone models or OS versions generate an enormous number of combinations [6]. The issue is significant for the Android platform because of its representatively large number of supported devices with different characteristics (e.g., screen resolution, size of memory, and set of sensors). The problem is compounded by the fact that a smartphone simulator or an emulator cannot fully substitute for the hardware [6]. At the same time, the development for different mobile platforms looks similar. Platforms have similar developer websites with necessary documentation, examples, and suggested patterns. The principles of the application life cycle are similar, for instance, comparing Android to the Windows Phone 7 [29].

Many software development companies are interested in the mobile market and many mobile platforms now exist: Android, iOS, Windows Phone, Symbian, etc. New ones appear regularly like the recent Ubuntu Mobile OS [30]. According to Gartner [31], Android devices have most of the market and Forbes says that the Android platform aims to meet enterprise requirements in the near future [32]. Previous research on the bug statistics for the Android OS [33] proved that the Android (with Symbian) has effectively organized an open-sourced bug-tracking system that deals with bugs and makes the platform better. The total number of applications in Google Play (www.appbrain.com/stats/) is now more than 850,000 and is increasing steadily. The open source nature of Android makes it popular among the scientific community, and many examples of research studies targeted at the Android system can be found.

4 Mobile Testing Services

To facilitate mobile testing, various cloud benefits are used and different TaaS, or supporting services, exist. Figure 1 provides references to them, along with mapping to correspondent testing stages. The presented types of testing were partially taken from a diagram on Perfecto Mobile's guide that shows the demanded device allocation during different Application Lifecycle Management (ALM) stages [27]. The diagram was extended by adding conceptualizations as a separate ALM activity, plus concept, security, and user experience (UX) testing, as well as highlighting test activities such as test planning, management, and issue tracking that are all specific to real-life mobile development.

The set of cloud services for mobile testing can be divided into three types: device clouds (mobile cloud platforms), services to support ALM, and tools to provide processing according to some testing techniques. The following sub-sections describe each type separately.

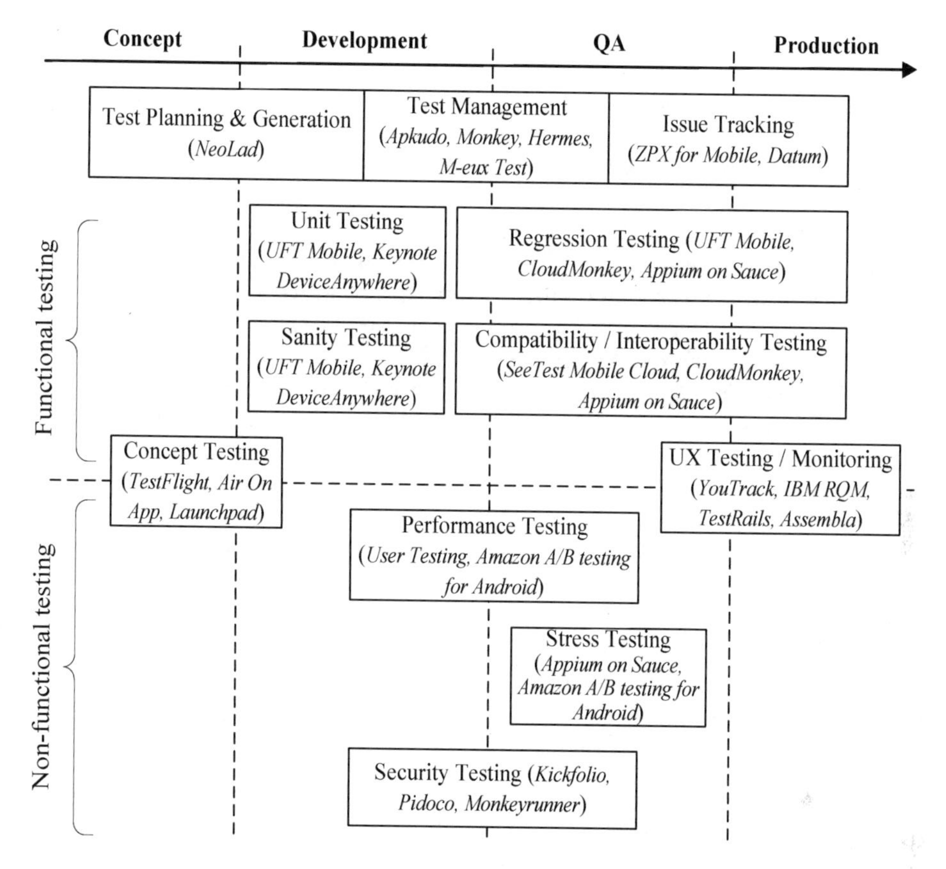

Fig. 1. Test Stages, Activities and Mobile Testing Services

4.1 Device Clouds

The majority of cloud services for mobile testing serves as a "cloud of devices" and provides remote access to smartphones in the cloud in order to accomplish testing, in other words, provides device hosting. Such services usually aid mobile developers in using remote smartphones as real devices for manual testing (interactive testing through a web interface), recording of scripts, and automatic running of tests on a range of models.

For instance, *Perfecto Mobile* service (www.perfectomobile.com) provides all of this functionality representing different modern hardware and software mobile platforms (Android, iOS, Windows Phone, and Symbian) and can be integrated with HP UFT (QTP) or MS Team Foundation Server. Devices available in the system have different parameters, for example, testing different types of Internet connections is possible. The service works with two kinds of test scripts: QTP and the Perfecto Mobile Application. Perfecto Mobile is only a public service, but UFT Mobile can also be deployed as a private cloud. UFT Mobile provides automated functional testing and special solutions for realistic mobile performance testing (e.g., LoadRunner and Performance Center).

Keynote DeviceAnywhere (www.keynotedeviceanywhere.com) is a similar service that provides online manual and automated testing of a mobile app on a variety of devices. It can be integrated with existing ALM through HP QTP, IBM RQM or special Java APIs.

The *SOASTA* service (www.soasta.com) provides two advanced solutions: TouchTest test automation for multi-touch, gesture-based applications and CloudTest for scalable mobile application testing (performance or load-testing with millions of geographically distributed emulated users). TouchTest scripts can be recorded and performed against user's own device. Users can control test devices via IP addresses.

The *Cigniti device cloud* (www.cigniti.com) provides remote access to a variety of mobile devices via own proprietary mobile test automation framework, with test accelerators for test automation and performance testing. Cigniti is suitable for network carrier testing.

SeeTest by Experitest (experitest.com) provides device cloud that can be deployed as a private platform within an organization. Test automation facilities include test script recording/performing on real devices or emulators and integration with HP UFT (QTP), TestComplete, C#, RFT, Java, Perl, Python. SeeTest also provides manual testing tools.

The *CloudMonkey* service (www.gorillalogic.com) runs MonkeyTalk scripts across many Android emulators and iOS simulators. Screenshot reports are positioned as the base testing results. CloudMonkey test jobs can be integrated with continuous integration (CI) servers like Jenkins.

The *Appium on Sauce* service (saucelabs.com) covers two functionalities: iOS device hosting and easy CI. The latter means that it can be used as a build server and testers do not need to set up developer environment on local machines. Test automation is implemented with Selenium, and interactive testing is only possible for web mobile applications. Appium can be deployed privately.

The *TestDroid* Cloud (testdroid.com) is a device cloud service oriented towards Android apps testing that uses the TestDroid AppCrawler engine to verify application devices' compatibility. TestDroid Recorder can be used to generate reusable Android JUnit test cases. Test results consist of screenshots and device logs. A tester can compare screenshots to check for GUI bugs. TestDroid can also be integrated with Jenkins or leveraged through REST APIs.

The *Scirocco* Cloud (www.scirocco-cloud.com) has all of the functionality of a device cloud, except of script recording. It supports only the Android platform and provides manual access to remote devices through its HTML5 web interface. Test automation is done by using one of three drivers: AndroidDriver, Monkeyrunner, or NativeDriver. Results are provided as a set of screenshots to compare.

The *LessPainfull* device cloud (www.lesspainful.com) is oriented for Android and iOS apps testing. As a test automation engine, it uses Calabash for Cucumber and accepts Cucumber-based test scripts. LessPainfull provides two options: private cloud tailored for single customer and shared cloud with devices common for several customers.

TestQuest (www.bsquare.com) is a distributed framework for deployment within an organization. It is oriented towards Android application testing and can be integrated with MS Visual Studio.

The *ZPX* service (www.zaptechnologies.com) provides device hosting and mobile test automation in the cloud and is compatible with HP ALM products.

Jamo (www.jamosolutions.com) provides a set of tools to perform remote and scheduled testing on a device. For instance, Wanconnector in combination with Remote Device Screen provides access to a device within different geographical locations. The M-eux Test tool supports web application testing.

Apkudo's Device Analytics (www.apkudo.com) provide some elements of multidirectional testing by testing devices (e.g., new smartphone models) against the top 200 apps from the market. Similar services are available for smartphone hardware testing, but these have no relation to mobile apps like Datum (www.metricowireless.com) that provides verification of calls, data quality, and video quality. Apkudo also offers free public and fully automated stress testing of the Android applications on the big range of models using the Monkey tool.

Table 1 summarizes the device clouds mentioned above and a comparison based on supported mobile platforms, types of testing, and delivery type of cloud solution.

Table 1. List of Device Clouds

Cloud Service	Supported Platforms			Types of Testing	Delivery Type	
	Android	iOS	Other		Public	Private
Apkudo	+			Stress (automated), New device approval	+	
Appium on Sauce		+		Manual for web applications, Automated	+	+
Cigniti	+	+	+	Automated, Interoperability, Performance, Network	+	
CloudMonkey	+	+		Automated, UI-oriented	+	+
DeviceAnywhere	+	+	+	Manual, Automated, Monitoring, Coverage	+	+
Jamo	+	+	+	Automated		+
Perfecto Mobile	+	+	+	Manual, Automated, Performance, Monitoring	+	
Scirocco Cloud	+			Manual, Automated	+	
SeeTest	+	+	+	Manual, Automated, On a new devices		+
SOASTA	+	+	+	Manual, Automated, Load, Performance, Gesture-based	+	+
TestDroid Cloud	+			Automated, UI-oriented, On a new devices	+	+
UFT Mobile	+	+	+	Automated, Load, Performance, Monitoring		+
Zap-Fix	+	+	+	Automated		+

Manual testing means the remote operation of a device via a web interface, and automated testing incorporates functional and regression testing and different kinds of automation. All device clouds provide compatibility testing as intended. Public cloud means service with shared devices, while a private cloud means an infrastructure allocated to a single user or a system to be deployed on a user-developer's site.

Two known research attempts within universities to create and investigate test-bed cloud solutions for mobile development are *SmartLab* [28] and the *Android Tactical Application Assessment and Knowledge* (ATAACK) Cloud [29]. Both are distributed systems that connect a set of mobile devices under the Android OS for application investigation, development, and testing.

The *SmartLab* is an experimental test-bed being developed at the University of Cyprus. It provides more than 40 connected Android smartphones plus emulated devices, but not many details are described or known.

The *ATAACK* Cloud is new joint project for Virginia Tech, the University of Maryland, and Vanderbilt University, with the support and funding by Air Force Research Laboratories. Its goal is large-scale mobile application testing and investigations.

These research studies consider device clouds with several smartphones connected to one computer (vertical) and several computers with connected smartphones (horizontal) scaling of devices, i.e., fully distributed systems, and how to provide access and testing.

Many studies regarding less-scaled test frameworks for distributed mobile testing [30] that are not cloud services and many tools for vertical-scaled test automation only exist, but their reviews are beyond the scope of this chapter.

All services mentioned in this section appear in Figure 3 with the following logistics: services that support the running of unit tests listed under "unit testing," services that support online manual testing listed under "sanity testing," references to script automation techniques of these services listed under "regression testing," all cloud devices listed under "interoperability/compatibility testing," and references to special integrated non-functional test approaches of these services.

4.2 Application Lifecycle Management Services

The application lifecycle management of mobile applications has own specifications and many cloud services exist that support test-related activities within ALM. Several examples of these cloud services are listed below.

1. Mobile developers, like all software developers, use issue tracking systems, e.g., with Agile-oriented plugins, more complex solutions like *IBM Rational Quality Manager*, or test management systems like *TestRails*. Some of these are integrated with software configuration management and facilitate code reviews or code style checks. A review of similar tools and solutions is not the goal of this work, so Figure 3 shows only several basic examples.
2. Mobile testing involves the use of actual hardware and so testers need additional knowledge and skills, such as build installation or crash-log retrieving. To facilitate beta build distribution activities, many cloud services exist. Some of them provide

functions for test team management (e.g. *HokeyApp*, hockeyapp.net) or build provisioning and deployment to the store (*AirOnApp* for iOS, www.aironapp.com). *TestFlight* service (testflightapp.com) helps to deal with the iOS build management and distributes them via email between separated testers. It provides an easy application installation on a real device, i.e., by a tap on the link in an email opened on a smartphone. A similar service for Android is *Launchpad* (launchpadapp.com). The *HokeyApp* (hockeyapp.net) build distribution provides extended functionality to collect live crash reports, feedback from users, and analysis of resulting test coverage. Usage of these services for build distribution can be integrated along with the continuous integration process of the company (e.g., via job scripts for the Jenkins build server).

3. User experience testing and monitoring of an app in production are required activities within mobile testing. Several analytics services gather usage statistics and these can be incorporated in a mobile application. *Perfecto Mobile* (www.perfectomobile.com) service also provides some solutions for monitoring performance. The following two services incorporate user experience testing in the build distribution facilities. The *UserTesting* service (www.usertesting.com) provides many real users who will examine an app and provide feedback about their experience with the app and thoughts about it. The *Amazon A/B testing for Android* (developer.amazon.com) provides a service that distributes two builds that differ in some features between two unique groups of users. Then it provides measurements and results about which feature is more successful.
4. Mobile development is very popular among startups and usually requires rapid prototyping for concept feasibility evaluation. Thus such services exist like *FluidUI* (www.fluidui.com) to easily create interactive prototypes, or *Kickfolio* (kickfolio.com) to share an app demo, or *Pidoco* (pidoco.com) to create realistic mockups. All of these are needed to test the concept and idea of the app (i.e., if it can hit the market) at a minimal expense.

4.3 Device Cloud-Based Testing Techniques

Device clouds provide different techniques for test automation (recording, distribution, and execution). This includes unit tests and GUI-based testing. Examples of approaches are standard Android SDK tools *Monkeyrunner* and *Monkey* (developer.android.com), special solutions like *SOASTA TouchTest*, and solutions based on object recognition (e.g., *Eggplant* automation based on VNC technology, www.testplant.com).

Test automation has its own weak sides, and according to experts in the field, cannot serve as a total substitution for manual testing. The issue that was noticed during the analysis of cloud test automation was the delivery of the test input data to mobile sensors (GPS, accelerometer, camera, etc.). While solutions to send dummy GPS coordinates exist, situation with a photo camera is more complicated because it requires the simultaneous changing of a picture (preferable physically in front of a camera) while performing a script. A variety of mobile apps use a camera as a part of their key functionality (e.g., shopping apps and QR code readers), and proper testing requires test cases with snapshots from different distances, angles, lights, etc. Other problematic aspects of automation are the sophisticated (approximate) screenshots comparisons, executions of direct device-to-device communication during the test, and others.

Device clouds provide compatibility, interoperability, and regression testing. Many services provide embedded tools to support performance monitoring and load testing (*Perfecto Mobile*, www.perfectomobile.com, *SOASTA*, www.soasta.com, *Cigniti Mobile Testing*, www.cigniti.com) or even automated stress testing on a variety of devices (e.g. *Apkudo*, www.apkudo.com).

There are special cloud services that aid with mobile performance and load testing. For instance, *SandStrom* (sandstorm.impetus.com) can be used for load testing of web mobile applications and *NeoLoad* (www.neotys.com) focuses on load testing of back-end servers by emulating typical mobile devices working in parallel and sending appropriate content to the server.

There are also standalone solutions for test techniques applications like performance frame counters on Windows Phone Emulator that theoretically can be leveraged in a cloud.

Security testing is mainly presented by static check techniques. *Checkmarks* (www.checkmarx.com) provides scanning of source code and supports Android and iOS applications. *Mobile App Security and Privacy Analysis* by Veracode (www.veracode.com) scans and evaluates binary files for vulnerabilities and can be leveraged through APIs.

Another type of services exists based on experts. For instance, *uTest* experts will assist with mobile security testing by manual penetration and using internal static and dynamic security testing solutions (www.utest.com). At the same time, research papers about novelty mobile security testing approaches exist (that potentially can be leveraged by some cloud services) [31], but they are out of the scope of this review.

Concept testing, UX testing, and monitoring techniques were comprehensively described in section 4.2 as parts of services that support ALM.

Mobile testing services should incorporate test planning and test generation techniques. *Keynote DeviceAnywhere Test Planner* (www.keynotedeviceanywhere.com) provides a coverage calculation for smartphone models to test that can be considered as application of combinatorial testing techniques, but it can be extended by using *pairwise*, *t-way*, or other approaches. *HokeyApp* only provides test coverage monitoring and analytics, i.e., the matrix of the devices and languages that were tested. *Cigniti Test Advisory Services* and *TestRails* provide more high-level test planning and control facilities.

The situation with cloud services for mobile testing is changing extremely rapidly: new ones appear and old ones get new functionalities. Thus, it is hard to guarantee that the provided list of tools and services is exhaustive, but it can serve as a useful baseline.

5 Standalone Tools for Mobile Application Testing

Any mobile platform has a correspondent software development kit (SDK) for app developers. Usually the producers of mobile platforms provide developers with a debugger, emulator or simulator, plugin for popular IDE, etc. The toolsets for Android, iOS, or Windows Phone development are very similar. Each platform also provides similar development support. For instance, web developer portals provide similar guidelines on how to use the available tools. In this section, we describe the most important standard tools (i.e., available from SDK) for Android app testing and several third-party extensions or analogues.

The basic tool for working with Android devices is *Android Debug Bridge* (ADB), which is a command-line utility to control Android devices. Device detection, debugging, execution of shell commands, and access to a device's file system is possible by using ADB. A high-level development environment like Eclipse (with the Android Development Tools plugin installed) implicitly uses ADB to install and debug builds within a connected device.

Android SDK provides two special tools for the GUI-based automated testing of applications. The first is *UI/Application Exerciser Monkey* (developer.android.com) for GUI stress testing, which generates a set of pseudo-random user events and sends them to an Android device. Previously, the *Apkudo* service (www.apkudo.com) was mentioned to provide a cloud of devices for long-term stress testing of an app using *Monkey*. It shows the statuses of the application being tested on each device, i.e., it either crashed after a sequence of random events or it is still running. Crash logs and other supporting information are provided.

A more advanced tool for automated testing provided by SDK is *Monkeyrunner* (developer.android.com), which runs on test scripts written in Python with several special classes available to provide support of touch, press, type, drag events, shell commands, intent invocations, app installations, and removal. Functionality is sufficient for basic GUI-based automation. So the following two strategies of interaction with interface components can be used:

(i) dynamic coordinates calculation (screen sizes can be dynamically retrieved);
(ii) and components enumeration through focus change.

At the same time a tester who writes test scripts should remember to put in appropriate delays (or special workarounds) between long-term events or actions and the results check. Monkeyrunner is suitable for screenshot analysis, as it provides methods to take screenshots during test script checkpoints and compare them. Thread-safeness is not guaranteed, but test scripts can include efficient simultaneous launches on several connected devices (and thus screenshots can be taken from several smartphones at the same time).

An *AndroidViewClient* extension (can be downloaded from Github) exists for Monkeyrunner that enables more high-level test scripts, particularly to address UI components in a test script by name or text. But this library only supports "rooted" devices with ViewServer installed or newer devices with Android's *UIAutomator* (Android API 16 and greater). *UIAutomator* is part of the Android SDK revision 21 and up and comes with the UIAutomatorViewer tool that lists all the UI objects.

Robotium (code.google.com) is another popular engine for the automated testing of Android applications. It is an extension of the Android test framework (JUnit tests for Androip applications) used to write easy and powerful automatic black-box tests.

Similarly, the Robolectric is based on JUnit 4 and runs Android tests directly on the JVM. Both of these tools point to another direction, i.e., the application of unit tests for mobile testing and even GUI-testing.

Other test automation solutions exist. Previously, several cloud services that provide a run of tests on multiple real devices were mentioned as having their own solutions for test automation. For instance, *LessPainful* (www.lesspainful.com) accepts test scripts written in Cucumber using *Calabash-Android* (github.com/calabash/). All of the aforementioned test automation drivers can be used for cloud-based testing of mobile systems. One of considered enhancements is to provide users with a choice of test scripts to use. The principles of usage are similar to Monkeyrunner, so it does not require a lot of work to integrate another driver like Robotium.

6 Research Studies in Mobile Testing

In the Table 2 we summarize recent research studies in the field of mobile testing. Each of them concerns a testing aspect that can be used in the cloud. For instance, many of the research studies deal with test automation, and theoretically, any service like device clouds can use described approaches as the test automation driver. In the same way, such extensions like test generation or static analysis can serve as an additional functionality integrated within any cloud service to facilitate mobile testing. Table 2 shows research areas and contributions for papers and highlights the year of release and the targeted mobile platforms. We can conclude that the popularity of mobile testing continues to grow and touches all possible aspects from effective test generation and design to execution and monitoring. At the same time, Android became the most popular platform under study. An open-source nature, prevalence in the market, support of an enormous number of devices, and ease of development (no provisions or jailbreaks are needed as in the case of iOS)—all make it the choice of researchers. These listed studies are potential directions for implementation of the integrated cloud services for mobile applications testing. They do not discuss cloud solutions for mobile testing, but instead present actual issues and techniques and describe possible supporting functionality.

7 Combinatorial Testing

Application of combinatorial approaches to mobile testing can aid in dealing with large amounts of different combinations of hardware and software parameters that should be covered by the tests. Coverage calculation is a crucial activity within mobile testing. So far, there are nine families of Android OS presented in the market (not counting lower sub-versions and correspondent builds without Google APIs), four types of screen resolutions (small, normal, large, and extra), and four levels of screen density.

Table 2. Researches in Mobile Testing

Year	Ref.	Mobile Platform	Research Area	Contribution
2012	[32]	Multi (J2ME)	Automation of mobile app testing	Framework that does not require a device under testing to be connected to a computer
	[31]	Android	Whitebox automated security testing of mobile apps	Fuzz test generation approach/testbed for emulation in the cloud
	[33]	Android	Automatic categorization of mobile apps	New method for categorizing Android applications through machine-learning techniques (while accepting malicious apps into the market)
	[34]	Android	GUI-based unit testing of mobile apps	Framework to test applications from GUI
	[35]	Android	Testing mobile apps through symbolic execution	Application of symbolic execution to generate test cases for mobile apps
	[36]	Android	Verification of touch screen devices	Test environment and supporting Android app to test touch screens
	[37]	Android	Automated mobile app testing through GUI-ripping	Technique and real-life case study of bug detection
2011	[38]	Android	GUI crawling-based testing of mobile apps	Technique for rapid crash testing and regression testing
	[39]	Multi	Model-driven approach for automating mobile app testing	Tool suite to apply Domain-Specific Modeling Language
	[40]	Android	Automation of mobile app testing	Review of the Android Instrume tation and the Positron frameworks
	[41]	Android	Automation of mobile app testing	Approach to use the Monkey tool in conjunction with JUnit
	[42]	Android (Dalvik)	Automated privacy testing of mobile apps	Automated privacy validation system to analyze apps (while they are accepted into the market)
	[43]	Multi (Android)	Automation of service-oriented mobile app testing	Approach for decentralized testing automation and test distribution
	[44]	Android	Model-based GUI testing of mobile apps	Extensive case study
	[45]	Multi	Automated test case design strategies for mobile apps	Comprehensive review of challenges and correspondent techniques
	[46]	Android	Static analysis of mobile apps	Extensions to Julia to provide formally correct analysis of mobile apps
	[47]	Android	GUI unit-testing of mobile apps	Techniques to assess the validity of the GUI code
2010	[48]	Multi (Android)	Adaptive random testing of mobile apps	Test case generation technique
2009	[49]	Windows Mobile	Automated GUI stress testing of mobile apps	Review/automated GUI stress testing tool
	[50]	J2ME	Automation of mobile app testing	Tool for testing mobile device applications
	[51]	Multi	Automation of mobile app testing	SOA based framework for mobile app testing

Other parameters like type of Internet connection (WiFi, 3G, or 4G), size of RAM, vendor, and a processor's characteristics should also be taken into account to provide adequate coverage during testing.

Many combinatorial testing materials can be found on the corresponding webpage of the National Institute of Standards and Technology (NIST) [52]. One of the simplest and easiest ways to implement combinatorial approaches is the Base Choice [53]. The idea is to create a base test case that represents the most important (common or popular) value for each parameter, and then create others by varying the value of only one parameter at a time. The base test case can be created using statistics, especially in case of mobile testing (i.e., what screen resolution is the most spread or what vendor shares the best part of the market). Pair-wise [54] and t-wise (t-way) [54] testing are the most common and powerful combinatorial testing approaches. According to the t-wise testing approach, for each subset of t input parameters of a system, every combination of valid values of these parameters should be covered by at least one test case. In pair-wise testing, which is a case of t-wise testing with t equals 2. The idea behind the t-wise approach is that the faults in the software are more likely triggered by a small number of input parameters, with the benefits being that t-wise testing providing reasonable coverage of software input space while using a small number of test cases. For example, if there are 15 Boolean input variables, the total number of various input combinations is 215 or 32,768. However, it takes only 10 input combinations (as pair-wise test cases) to cover all of the different values for each pair of input variables.

Some examples of combinatorial tests based on different configurations of Android application can be found in [56]. Other similar techniques, including t-wise testing [57], MC/DC [58], and RC/DC [59] testing criteria are also worth to be mentioned. The *ACTS tool* (csrc.nist.gov) created by the NIST and the *ALLPAIRS* (www.satisfice.com) provide engines to calculate different combinatorial strategies and perform combinatorial testing.

8 Conclusions

Ensuring quality of modern mobile applications is complicated by a variety of mobile hardware and software platforms, variety of sensors, network interfaces, existence of web mobile and hybrid applications, and also high user's expectations. This is why thorough testing of mobile applications is of a great importance for both developers and consumers of these products.

Nowadays, testing extensively migrates to the clouds allowing to support team work, shorten testing time, and to reduce development costs, that is especially important for many startup companies. In the paper we have described a set of cloud services for mobile testing that can be divided into three types: (i) device clouds (mobile cloud platforms), (ii) services to support application lifecycle management, and (iii) tools to provide processing according to some testing techniques. Mobile testing over a cloud is an extremely important activity that is very hard to research. As it was described above, a lot of cloud services exist that fulfill the initial testers' needs, but a scalable platform for effective crowdsourcing in mobile testing supporting multidirectional testing and flexible integration of many different testing services and techniques is still of a great demand.

References

1. White, J., Clarke, S., Doughtery, B., Thompson, C., Shmidt, D.: R&D Challenges and Solutions for Mobile Cyber-Physical Applications and Supporting Internet Services. Springer Journal of Internet Services and Applications 1(1), 45–56 (2010)
2. Work, D.B., Bayen, A.M.: Impacts of the Mobile Internet on Transportation Cyberphysical Systems: Traffic Monitoring using Smartphones. In: National Workshop for Research on High-Confidence Transportation Cyber-Physical Systems: Automotive, Aviation and Rail, Washington, DC, USA (2008)
3. Leijdekkers, P., Gay, V.: Personal Heart Monitoring and Rehabilitation System using Smart Phones. In: Intern. Conf. on Mobile Business, Copenhagen, Denmark (2006)
4. Moser, K.: Improving Work Processes for Nuclear Plants. In: American Nuclear Society Utility Working Conf., Hollywood, Florida, USA (2012)
5. Wasserman, A.: Software engineering issues for mobile application development. In: Workshop on Future of Software Engineering Research at the 18th Int. Symposium on Foundations of Software Engineering (ACM SIGSOFT), Santa Fe, USA, pp. 397–400 (2010)
6. The Essential Guide to Mobile App Testing, `http://www.utest.com/landing-blog/essential-guide-mobile-app-testing`
7. Holler, R.: Mobile Application Development: A Natural Fit with Agile Methodologies, `http://www.versionone.com/pdf/mobiledevelopment.pdf`
8. Vilkomir, S.: Cloud Testing: A State-of-the-Art Review. Information & Security: An International Journal 28(2(17), 213–222 (2012)
9. Tilley, S., Parveen, T.: Software Testing in the Cloud: Perspectives on an Emerging Discipline. IGI Global (2012)
10. Tsai, W., Chen, X., Liu, L., Zhao, Y., Tang, L., Zhao, W.: Testing as a service over cloud. In: 5th IEEE Int. Symposium on Service Oriented System Engineering, pp. 181–188 (2010)
11. Kalliosaari, L., Taipale, O., Smolander, K.: Testing in the Cloud: Exploring the Practice. IEEE Software 29(2), 46–51 (2012)
12. Weidong, F., Yong, X.: Cloud testing: The next generation test technology. In: 10th Int. Conf. Electronic Measurement & Instruments, Chengdu, China, pp. 291–295 (2011)
13. Inçki, K., Ari, I., Soze, H.: A Survey of Software Testing in the Cloud. In: IEEE 6th Int. Conf. on Software Security and Reliability Companion, pp. 18–23 (2012)
14. Priyanka, C.I., Rana, A.: Empirical evaluation of cloud-based testing techniques: a systematic review. ACM SIGSOFT Software Engineering Notes Archive 37(3), 1–9 (2012)
15. Mote, D.: Cloud based Testing Mobile Apps. In: 2nd IndicThreads.com Conference on Software Quality, Pune, India (2011)
16. Cloud Testing: Database of Cyber Security and Information Systems Information Analysis Center, `https://sw.thecsiac.com/databases/url/key/7848/8764/8765#.USGPb-h8vDm`
17. Tilley, S., Parveen, T.: Software Testing in the Cloud: Migration & Execution. Springer Briefs in Computer Science (2012)
18. Riungu, L.M., Taipale, O., Smolander, K.: Research Issues for Software Testing in the Cloud. In: IEEE 2nd Int. Conf. Cloud Computing Technology and Science, pp. 557–564 (2010)
19. Rhoton, J., Haukioja, R.: Cloud Computing Architected: Solution Design Handbook. Recursive (2011)

20. Coulouris, G., Dollimore, J., Kindberg, T., Blair, G.: Distributed Systems: Concepts and Design. Addison-Wesley (2011)
21. Muccini, H., Francesco, A., Esposito, P.: Software testing of mobile applications: Challenges and future research directions. In: 7th Int. Workshop on Automation of Software Test (2012)
22. Franke, D., Weise, C.: Providing a Software Quality Framework for Testing of Mobile Applications. In: IEEE 4th Int. Conf. on Software Testing, Verification and Validation, Berlin, Germany, pp. 431–434 (2011)
23. Milano, D.: Android Application Testing Guide. Publishing Ltd. (2011)
24. Frederick, G., Lal, R.: Testing a Mobile Web Site. Beginning Smartphone App Development – Part IV. Apress (2009)
25. Dantas, V., Marinho, F., Da Costa, A., Andrade, R.: Testing requirements for mobile applications. In: 24th Int. Symposium on Computer and Information Sciences (2009)
26. Test Strategies for Smartphones and Mobile Devices, `http://www.macadamian.com/images/uploads/whitepapers/MobileTestStrategies_Aug2010.pdf`
27. Make Your Mobile Testing Solution Enterprise-Ready, `http://www.perfectomobile.com/portal/cms/resources/enterprise-ready_white-paper`
28. Konstantinidis, A., Costa, C., Larkou, G., Zeinalipour-Yazti, D.: Demo: a programming cloud of smartphones. In: 10th Int. Conf. on Mobile Systems, Applications, and Services, pp. 465–466 (2012)
29. Turner, H., White, J., Reed, J., Galindo, J., Porter, A., Marathe, M., Vullikanti, A., Gokhale, A.: Building a Cloud-Based Mobile Application Testbed. IGI Global (2012)
30. She, S., Sivapalan, S., Warren, I.: Hermes: A Tool for Testing Mobile Device Applications. In: Software Engineering Conf., Queensland, Australia (20091)
31. Mahmood, R., Esfahani, N., Kacem, T., Mirzaei, N., Malek, S., Stavrou, A.: A whitebox approach for automated security testing of Android applications on the cloud. In: 7th Int. Workshop on Automation of Software Test, pp. 22–28 (2012)
32. Nagowah, L., Sowamber, G.: A Novel Approach of Automation Testing on Mobile Devices. Int. Conf. on Computer & Information Science 2, 924–930 (2012)
33. Sanz, B., Santos, I., Laorden, C., Ugarte-Pedrero, X., Bringas, P.: On the Automatic Categorisation of Android Applications. In: 9th Annual IEEE Consumer Communications and Networking Conf. – Security and Content Protections (2012)
34. Allevato, A., Edwards, S.: RoboLIFT: simple GUI-based unit testing of student-written android applications. In: 43rd ACM Technical Symposium on Computer Science Education, p. 670 (2012)
35. Mirzaei, N., Malek, S., Păsăreanu, C., Esfahani, N., Mahmood, R.: Testing Android Apps Through Symbolic Execution. ACM SIGSOFT Software Engineering Notes Archive 37(6), 1–5 (2012)
36. Zivkov, D.: Touch screen mobile application as part of testing and verification system. In: 35th Int. Convention MIPRO, pp. 892–895 (2012)
37. Amalfitano, D., Fasolino, A., Tramontana, P., De Carmine, S.: Using GUI ripping for automated testing of Android application. In: 27th IEEE/ACM Int. Conf. on Automated Software Engineering, Germany (2012)
38. Amalfitano, D., Fasolino, A.R., Tramontana, P.: A GUI Crawling-Based Technique for Android Mobile Application Testing. In: Software Testing, Verification and Validation Workshops, pp. 252–261 (2011)

39. Ridene, Y., Barbier, F.: A model-driven approach for automating mobile applications testing. In: 5th European Conf. on Software Architecture: Companion (2011)
40. Kropp, M., Morales, P.: Automated GUI Testing on the Android Platform. IMVS Fokus Report 4(1) (2010)
41. Hu, C., Neamtiu, I.: Automating GUI testing for Android applications. In: 6th Int. Workshop on Automation of Software Test, pp. 77–83 (2011)
42. Gilbert, P., Chun, B., Cox, L., Jung, J.: Automating Privacy Testing of Smartphone Applications. Technical Report CS-2011-02 (2011)
43. Edmondson, J., Gokhale, A., Neema, S.: Automating Testing of Service-oriented Mobile Applications with Distributed Knowledge and Reasoning. In: Service-Oriented Computing and Applications, pp. 1–4 (2011)
44. Takala, T., Katara, M., Harty, J.: Experiences of System-Level Model-Based GUI Testing of an Android Application. In: Software Testing, Verification and Validation, pp. 377–386 (2011)
45. Selvam, R., Karthikeyani, V.: Mobile Software Testing – Automated Test Case Design Strategies. Int. J. on Computer Science and Engineering (2011)
46. Payet, É., Spoto, F.: Static Analysis of Android Programs. In: Bjørner, N., Sofronie-Stokkermans, V. (eds.) CADE 2011. LNCS, vol. 6803, pp. 439–445. Springer, Heidelberg (2011)
47. Sadeh, B., Ørbekk, K., Eide, M.M., Gjerde, N.C.A., Tønnesland, T.A., Gopalakrishnan, S.: Towards Unit Testing of User Interface Code for Android Mobile Applications. In: Zain, J.M., Wan Mohd, W.M.b., El-Qawasmeh, E. (eds.) ICSECS 2011, Part III. CCIS, vol. 181, pp. 163–175. Springer, Heidelberg (2011)
48. Liu, Z., Gao, X., Long, X.: Adaptive random testing of mobile application. In: Computer Engineering and Technology, pp. 297–301 (2010)
49. Abdallah, N., Ramakrishnan, S.: Automated Stress Testing of Windows Mobile GUI Applications. In: 20th Int. Symposium on Software Reliability Engineering (2009)
50. Sivapalan, S., Warren, I.: Hermes: A Tool for Testing Mobile Device Applications. In: Software Engineering Conference, Australia (2009)
51. Liu, Z.-F., Liu, B., Gao, X.-P.: SOA based mobile application software test framework. In: 8th Int. Conf. Reliability, Maintainability and Safety, pp. 765–769 (2009)
52. Combinatorial Methods in Software Testing, http://csrc.nist.gov/groups/SNS/acts/
53. Grindal, M., Offutt, J., Andler, S.F.: Combination Testing Strategies: a Survey. Software Testing, Verification and Reliability 15(3), 167–199 (2005)
54. Kuhn, D.R., Lei, Y., Kacker, R.: Practical Combinatorial Testing - Beyond Pairwise. IEEE IT Professional 6, 19–23 (2008)
55. Maximoff, J.R., Trela, M.D., Kuhn, D.R., Kacker, R.: A Method for Analyzing System State-space Coverage within a t-Wise Testing Framework. In: IEEE Int. Systems Conf., San Diego (2010)
56. Kuhn, D.R., Kacker, R.N., Lei, Y.: Practical Combinatorial Testing. NIST Special Publication 10, 13–15 (2010)
57. Lei, Y., Kacker, R., Kuhn, D.R., Okun, V., Lawrence, J.: IPOG: A General Strategy for T-Way Software Testing. In: IEEE Engineering of Computer Based Systems Conf., pp. 549–556 (2007)
58. Chilenski, J.J., Miller, S.: Applicability of Modified Condition/Decision Coverage to Software Testing. Software Engineering J. 9, 193–200 (1994)
59. Vilkomir, S., Bowen, J.P.: From MC/DC to RC/DC: Formalization and Analysis of Control-flow Testing Criteria. Formal Aspects of Computing 18(1), 42–62 (2006)

Agent Approach to Network Systems Dependability Analysis in Case of Critical Situations

Jacek Mazurkiewicz

Institute of Computer Engineering, Control and Robotics, Wroclaw University of Technology
ul. Janiszewskiego 11/17, 50-372 Wroclaw, Poland
Jacek.Mazurkiewicz@pwr.wroc.pl

Abstract. The chapter describes the analysis and discussion of the network systems in case of the critical situation that happens during ordinary work. The formal model is proposed – based on the two types of real sophisticated network systems – with the approach to its modeling based on the system behavior observation. The agent approach to constant network monitoring is given using hierarchical structure. The definition of the critical situation sets are created by reliability, functional and human reasons. The proposed method is based on specified description languages that can be seen as a bridge between system description and an analysis tools. Using a multilevel-agent based architecture the realistic data are collected. Described architecture can be finding as a basis for a tool that can visualize and analyze data, with respect to real parameters. No restriction on the system structure and on a kind of distribution describing the system functional and reliability parameters is the main advantage of the approach. The proposed solution seems to be essential for the owner and administrator of the transportation systems.

Keywords: network systems, critical sets, reliability, dependability modeling.

1 Introduction

Contemporary network systems are very often considered as a set of services realized in well-defined environment created by the necessary hardware and software utensils. The system dependability can be described by such attributes as *availability* (readiness for correct service), *reliability* (continuity of correct service), *safety* (absence of catastrophic consequences on the users and the environment), *security* (availability of the system only for authorized users), *confidentiality* (absence of unauthorized disclosure of information), *integrity* (absence of improper system state alterations) and *maintainability* (ability to undergo repairs and modifications) [1, 3, 13, 20].

The system realizes some tasks and it is assumed that the main system goal, taken into consideration during design and operation, is to fulfill the user requirements. The system functionalities (services) and the technical resources are engaged for task realization. Each task needs a fixed list of services which are processed based on the system technological infrastructure. The different services may be realized using the same technical resources and the same services may be realized involving different sets of the

W. Zamojski and J. Sugier (eds.), *Dependability Problems of Complex Information Systems*,
Advances in Intelligent Systems and Computing 307, DOI: 10.1007/978-3-319-08964-5_5

technical resources. It is easy to understand that the different values of performance and reliability parameters should be taken into account. The last statement is essential when tasks are realized in the real system surrounded by unfriendly environment that may be a source of threads and even intentional attacks.

Moreover, the real systems are built on the base of unreliable technical infrastructures and components. The modern systems are equipped with suitable measures and probes, which minimize the negative effects of these inefficiencies (a check-diagnostic complex, fault recovery, information renewal, time and hardware redundancy, reconfiguration or graceful degradation, restart etc). The contemporary network systems are created as very sophisticated products of human idea characterized by the complex structure. This way the critical situations observable during its exploitation are not always predictable for system owners and managers, but could be very costly for a company and sometimes even damaging.

The necessary analysis mechanisms should be created not only for the money saving, by also as a tool for the future administration of the system and decision support (based on some specified metrics). The main problem is to realize multi-criteria optimization for system management. The solution ought to combine the sets of reliability, functional and economic parameters. The mentioned data are modeled by distributions - so it makes the optimization problem more sophisticated. This is the reason why we propose the computational collective intelligence to create the device to support human's decisions.

The presented work uses the agents in task of the transportation system monitoring and modeling, so we propose the following description of the most important agent's features [5]:

- unique identification within the proposed architecture,
- interaction abilities and proper interfaces for communication and different data transfer,
- secure protocols necessary for communication purposes,
- hardware and / or software implementation,
- plug-and-play ability to guarantee promising scalable and flexible structure.

The temporary computer engineering still does not define an "agent" term in detailed way, but it is not a real barrier to establish the unified semantic meaning of the word in technical point of view. The agent can play the role of the autonomous entity [6] as a model or software component for example. The agent's behavior can be noticed as trivial reactions, but is not limited – so we can easily find agents characterized by complex adaptive intelligence. Sometimes is important to point the potential adaptive abilities of the agents [7]. It means the agent can gather the knowledge from the environment around and to tune their behavior as a reaction for different events.

This way we can say the agents belong to the softcomputing world. The agent's structure is not obligatory plain. We can easily [5] find at least two levels (lower, higher) of the rules created for the agents. This approach allows to tune the level of the sensitivity for the environment and to define the vitality feature of the agent understood as activity or passivity [12, 15].

The agent-based approach provides the real great effectiveness comparing with the classical architectures if we think about the data gathering and aggregation from the real sophisticated system characterized by the large network, significant number of nodes and non-trivial addressing aspects. This way it is easy to create the global and detailed enough view for multilevel systems with elements described by various sets of features. We propose to use the agents to create the intelligent hierarchical monitoring architecture - described in section 4. The section 5 presents a solution of a description language for a proposed model, called *SML* (*System Modeling Language*). As a format of the proposed language *XML* (*Extensible Markup Language*) was chosen. Main reason is a simple (easy to learn) and readable structure, that can be easily convert to text or other format. Moreover, *XML* is supported not only with various tools (providing validation possibilities) but is also supported by many programming languages and framework in case of quicker and more efficient implementation.

The aim of this chapter is to point the problems of the critical situations in unified network system – product of essential elements and features taken from two kind of real systems: *Discrete Transport System (DTS)* and *Computer Information System (CIS)*. Each part of the system is characterized by unique set of features and can caused the critical situation of whole system if it starts to work in unusual way or the fault or error of it is noticed. It is hard for an administrator, manager or an owner to understand the system behavior and to combine the large scale of variant states of it in single – easily observable and controlled global metric as a pointer to make the proper decision in short time period. To overcome this problem we propose a functional approach. The system is analyzed from the functional point of view, focusing on business service realized by a system [21]. The analysis is following a classical [15]: modeling and simulation approach. It allows calculating different system measures, which could be a base for decisions related to administration of the transportation systems. The results of the system observation – understand as the set of data collected during the simulation process are the basis to define the critical situations and they allow providing proper solution to lift-up the systems in effective way if the critical situation occurs. This is the only sensible way, because the critical situations are the real and not removable part of the system life. The organization of this paper is as follow. We start with description of the abstract service network model (section 2). Base in it we define the normal conditions of the system work (section 3). In section 6 we provide the most adequate – in case of the level of detail - the well-established description of the critical situation.

2 Network Model

The chapter describes approach based on functional-dependability models understood as a concept of specifying dependability aspect for two perspectives: secure and dependable system as much as service-related operational system. In our research, we focus on two types of service models, that where close to our interest area: *Discrete Transport System* (*DTS*) [16, 17, 20, 21, 22] and *Computer Information System* (*CIS*) [12, 13, 17, 18]. Both systems can be analyzed separately, but because of their specific goal, some common mechanisms can be seen. Taking into consideration more generic perspective, we decided to focus on a common view on the system model we call - *Abstract Service Network Model*.

As mentioned, both systems have the same aim – to provide a service in a sense of user request accomplishment. For this reasons, the key point to analyze the systems is a *Task* (*T*) given to the systems. Task is defined by the user and parameters related with time (user patience time, delivery take, etc.) but also it is strongly and inextricably connected with some service scenario. In fact, when we analyze logically the way the service is provided, we can see that the scenario conditions define specific choreography (graph of various components) within a service. The choreography must be defined and known. Since task is realized as an input to the *Business Service* (*BS*), therefore its choreography is based on predefined service components located in network nodes (reconfigurable components).

Moreover, network nodes base on *Technical Infrastructure* (*TI*) – resources used as elements for providing dependable service seen as a hardware and software linked within a network. Various functional define each element of the Technical Infrastructure (routes and central points in *Discrete Transport System*, computers or network devices in *Complex Information Systems*) and dependability parameters, not to mention about some time functions. Time related with the technical resources is as much important as time on a service level, therefore we speak about – *Chronicle of the System* (*K*). Taking into consideration these common features an abstract model can be proposed as follows:

$$ANS = < T, BS, TI, M, K > \tag{1}$$

where: *ANS* – Abstract Network Services, *T* – Task, *BS* – Business Service, *TI* – Technical Infrastructure, *M* – User, *K* – Chronicle of the System.

The unified description can guarantee the required level of abstraction for the analysis we are going to provide.

3 Service Description

3.1 Tasks

The problems of the contemporary systems reliability certainly need to be extended to cover the envisaged fact that the main object (system) of its studies is a tightly connected complex of hardware resources, information resources (algorithms and procedures of operations and system management) and human-factor (managers, administrators and users). The studied systems realize complex functions and are capable of substituting tasks on detecting faults (functional redundancy). The systems operate in a changing environment, often antagonistic to them. Users generate tasks which are being realized by the system. The task to be realized requires some services (functionalities) available in the system. A realization of the service needs a defined set of technical resources. In a case when any resource component of this set is in a state "out of order" or "busy", the task may wait until a moment when the resource component returns to a state "available" or the service may try to create other configuration based on available technical infrastructure [2, 3, 4].

A technological infrastructure is considered as a set of hardware resources (devices and communication channels) which are described by sets of their technological, reliability and maintenance parameters. The information resources are understood in the same way. The human-factor's functions are defined little bit different: she or he can be defined as: a system operator, a service person, a system manager (administrator) etc. [21, 29]. The system management allocates the resources to the task realisation, checks the efficient states of the system, performs the suitable actions to locate faults, attacks or viruses and to minimise their negative effects. In many situations the system staff and the management system have to cooperate in looking for adequate decisions (for instance to fight with a heavy attack or when a new virus is disclosed). The system events corresponds to: tasks realisation, occurrence of incidents (faults, viruses, attacks) and system reactions to them (technological and information renewals). Task configurations change when the tasks are being processed. The software management, reacting with the system users, determines the changes. Some changes may be the result of detecting system faults and reacting to them. This is called system reconfiguration [25, 27]. The subsets of resources used by the tasks do not need to be disjoint. A resource that can be allocated to more than one configuration at the same time is called sharable, whereas one that cannot is non-sharable. Some resources, for example the central processors in computer systems, are "time-sharable". This is a technique that allows sharing of resources that are essentially non-sharable, by very fast switching of the allocation in time [1, 17, 18].

3.2 Events

Different events of the service network are considered as: *normal* functional events described by such time parameters as the start or / and the end of the task, a moment of a system resources allocation, a time of occurrence of a new task, an (prognoses or real) task execution time, etc., *unfriendly* incidents that are disturbed efficient system execution; for example failures of transport structure, failures and errors, delay time of data packages, faults of network devices or dispatching system, etc.

It is easy to notice that the first class of system events is strictly connected with correct system task realization and the second one groups all events disrupting the efficient operation of the system and which may start the system defence reactions. In this way the first class of events will be called "efficient functional events" and the second one "dependable incidents" or "unfriendly events". A classification of dependable incidents and system reactions is presented in the Figure 1. A dependable incident is an event that might lead to some disruptions in the system behavior. The incident may cause some damage to the system resources; transport structure, management actions and, in consequence, it may disrupt the executed transport processes [3, 4]. If a fault appears during the task execution then the system on the base of decision of its management system starts renewal processes. Time of technological renewal activities are added to the nominal time of the task so a real time of the task duration will be longer. The real duration time of the executed tasks depends on the nature of the system faults. Failures of hardware may need both renewals of technological resources and information resources. Consequences of human errors or computer software faults are limited to renewals of information processes. Sometimes an incident which are occurred in a short time interval may have a more serious impact on the system behavior; it may escalate to a security incident, a crisis or a catastrophe. The failures of the network structure - physical failures of technical infrastructure need to use adequate service teams, spare elements or substituted routes. Very often "technical" system renewal processes are considered with assuming of the limited resources, for example the number service team for the part of the network [5, 27].

Other sources of the network disruptions we can find in organization and management: overloading of the technical infrastructure, traffic problems or jams – caused by limited bandwidth or dispatching errors, dispatching faults – system is not able to keep up the dynamic changes of the situation in the working network.

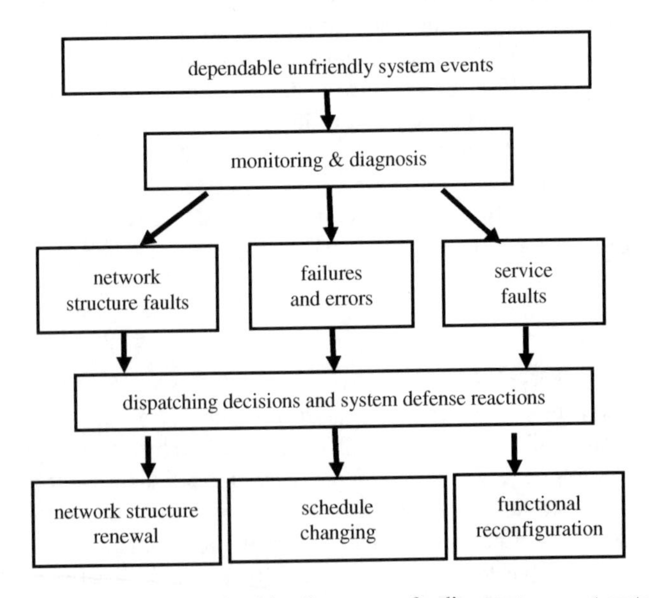

Fig. 1. A classification of unfriendly events of a discrete transport system

In these cases exploitation system renewal processes are initiated by the system dispatcher. The processes very often consume more time and money than a renewal of a "simple (physical)" broken technical resource, e.g. a repair of a failed truck or a lift.

3.3 System Maintenance

The modern systems are equipped with suitable measures, which minimise the negative effects of these inefficiencies (a check-diagnostic complex, fault recovery, information renewal, time and hardware redundancy, reconfiguration or graceful degradation, restart etc). The special services resources (service persons, different redundancy devices, etc.) supported by the so-called maintenance policies (procedures of the service resources using in purpose to minimise negative consequences of faults that are prepared before or created ad hoc by the system manager) are build in every real system [3, 4, 24, 27]. The maintenance policy is based on two main concepts: detection of unfriendly events and system responses to them. Detection mechanisms should ensure detection of incidents based on observation of a combination of seemingly unrelated events, or on an abnormal behaviour of the system. Response provides a framework for counter-measure initiatives to respond in a quick and appropriate way to detected incidents. In general, the system responses incorporate the following procedures: detection of incidents and identification of them, isolation of damaged resources in order to limit proliferation of incident consequences, renewal of damaged processes and resources. Relations among the incidents and the reactions of the system are shown in Figure 2. A services network is a system of functional services that are necessary for clients tasks realisation process. The services networks are organized based on the technical infrastructure and technological services which are involved into a task realisation process according to decisions of the management system. The task realisation process may include many sequences of services, functions and operations which are using assignment network resources. Description of the allocation of network services and their implementation process will be hereinafter referred to as network choreography. We can build more general definition of the system introducing the idea of the net of services. It is described at the upper level of abstraction: a task or a job may use a single service or a few services - concurrent or sequenced - on the base of available network resources. The management system - allocates services (functionalities) and network resources to realized tasks, checks states of the services network and controls suitable system responses to detected and localized unfriendly events and minimizes their negative effects. The control of the defence reactions of the system is understood as the choice of an appropriate maintenance policy. A service may be realised based on a few separated sets of functionalities with different costs which are the consequences of using different network resources. Because the services have to cooperate with other services than protocols and interfaces between services, and/or individual activities are crucial problems which have a big impact on the definitions of the services, and on processes of their execution. Generally the management system has main functionalities: monitoring of network states and controlling of services and resources; creating and implementing maintenance policies which ought to be adequate network reactions on concrete events/accidents. In many critical situations a team of persons and the management system have to cooperate in looking for adequate counter-measures. As a consequence, the services network is considered as a dynamical structure with many streams of events generated by realized tasks, used services and resources.

Some network events may been independent but majority of events depends on a history of a network life. Generally, event streams created by a real network are a mix of deterministic and stochastic streams which are strongly tied together by a network choreography. Modelling of this kind of systems is a hard problem for system designers, constructors and maintenance organizers, and for mathematicians, too. It is proposed to focus the dependability analysis of the networks on the fulfilment of requirements defined by user task [27].

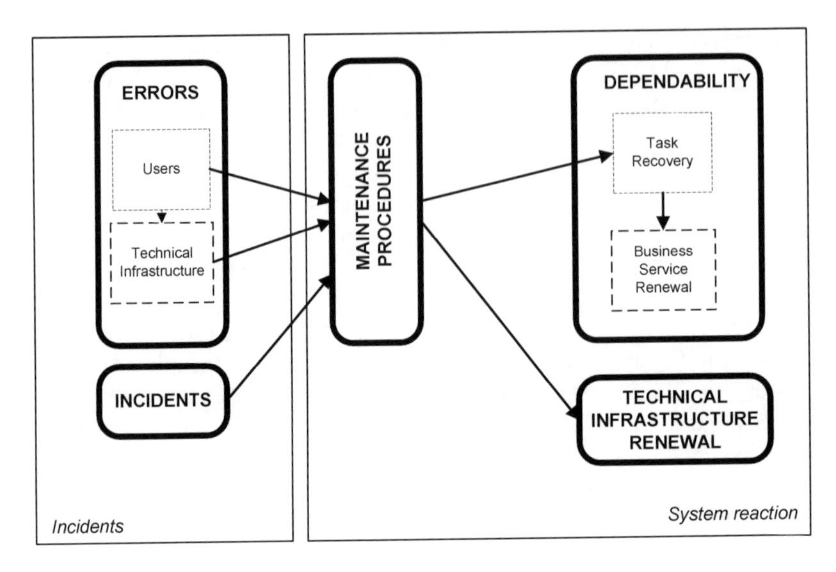

Fig. 2. Incidents and reactions of the system

Therefore, it should take into consideration following aspects:

- specification of the user requirements described by task demands, for example expected volume to transport, desired time parameters etc.,
- functional and performance properties of the network system and components,
- reliable properties of the technical infrastructure that means reliable properties of the network structure and its components considered as a source of unfriendly events which influence the task processing,
- threads in the network environment,
- measures and methods which are planned or build-in the network system for elimination or limitation of unfriendly incident consequences; reconfiguration of the transport system is a good example of such methods,
- the system of maintenance policies applied in the considered network.

The task realisation process is supported by two-level decision procedures connected with selection and allocation of the network services (functionalities) and infrastructure resources. The first level of decision procedure is focused on suitable services selection and a task configuration. The functional and the performance task demands are based on suitable services choosing from all possible network services. The goal of the second level of the decision process is to find needed components of the network infrastructure for each service execution and the next to allocate them based on their availability to the service configuration. If any component of technical infrastructure is not ready to support the service configuration then the allocation process of network infrastructure is repeated. If the management system could not create the service configuration then the service management process is started again and other task configuration may be appointed. These two decision processes are working in a loop which is started up as a reaction on network events and incidents [3, 4, 8, 27].

3.4 Dependability Discussion

The term dependability is well known in the literature and commonly used by fault tolerant and dependable computing community, but has been assigned many different meanings. For example, there is more than one definition of dependability [3, 4, 9, 10, 14]. The dependability of the system can be defined as the ability to execute the functions (tasks, jobs) correctly, in the anticipated time, in the assumed work conditions, and in the presence of threats, technological resources failures, information resources and human faults (mainly malfunctions) [8]. Dependability is the most comprehensive concept for modeling complex systems taking a top-down approach [1]. It is evolving into a distinct discipline attempting to subsume the preceding concepts of reliability, and fault-tolerance. There is no universally accepted definition of dependability; the term has been accepted for use in a generic sense as an umbrella concept [2, 3].

Users of the system realize some tasks using it – for example: send a parcel in the transport system or buy a ticket in the internet ticket office. It is assumed that the main goal, taken into consideration during design and operation, is to fulfill the user's requirements. We can easy find some quantitative and qualitative parameters of user's tasks [2, 27]. The system functionalities (services) and the technical resources are engaged for task realization. Each task needs a fixed list of services, which are processed based on the system technological infrastructure or the part of it. The different services may be realized using the same technical resources and the same services may be realized involving different sets of the technical resources. It is easy to understand that the different values of performance and reliability parameters are taking into account. The last statement is essential when tasks are realized in the real system surrounded by unfriendly environment that may be a source of threads and even intentional attacks. Moreover, the real systems are built of unreliable software and hardware components as well.

It is hard to predict all incidents in the system; especially, it is not possible to envision all possible attacks, so system reactions are very often "improvised" by the system, by the administrator staff or even by expert panels specially created to find a solution for the existing situation. The time, needed for the renewal, depends on the incident that has occurred, the system resources that are available and the renewal policy that is applied. The renewal policy should be formulated on the basis of the required levels of system dependability (and safety) and on the economic conditions (first of all, the cost of downtime and lost processing/computations) [2]. Modeling of this kind of systems is a hard problem for system designers, constructors and maintenance organizers, as well as for mathematicians. It is worth to point out some achievements in the computer science area such as Service Oriented Architecture [3, 4, 26] or Business Oriented Architecture [26, 29], and a lot of languages for network description on a system choreography level, for example *WS-CDL* [18], or a technical infrastructure level, for example *SDL* [18, 27]. The approach seems to be useful for analysis of a network from the designer point of view. The description languages are supported by the simulation tools, for example modified *SSF Net* simulator [21, 22]. Still it is difficult to find the computer tools which are combination of model languages and Monte Carlo simulators [19, 23, 24].

4 Monitoring Architecture

In case of Monitoring Architecture representation and distributed multilevel agent-based architecture can be constructed. Figure 3 shows the diversification of complexity of a system into layers and their placement in a system. The lowest components of the structure are *Node Probes* (*NP*) which are the simplest pieces of the architecture representing resident level. These are the simplest and easy to get data that at this level represent small value that is why they are aggregated in upper units are forwarded to appropriate supervising *Node Sensors* (*NS*). Next *Node Sensors* collect the data and create an image of the particular area – so they are located in the ordinary nodes (*ON*). Again the information is sent to a higher level – *Local Agent* (*LA*) – combined to the central node (*CN*).

$$NS_i = \bigcup_j NP_j;\ j \in N \tag{2}$$

This set of information creates a database building representation of local part of a system (subnetwork). It means that the local view of the system and partial administration in the system can be done at this level.

$$LA_i = \bigcup_j NS_j;\ j \in N \tag{3}$$

The highest component of this structure is the *Global Agent* – working in the headquarter (*HQ*), that picks and process local information's and view to one central unit.

$$GA = \bigcup_j LA_j;\ j \in N \tag{4}$$

This module stores all information from a whole system. It is situated in one point and one dedicated machine (with a strong backup). Assembling all local view at this level we get one homogenous global view. At this level, data-mining techniques can be used. We can see that set of information flow goes to the central unit – *Global Agent*. For this reasons it is the most complex and the simplicity of the data that are needed to describe the system in this point is the highest in hierarchy.

5 Description Language

Since, the purpose of the work is to analyze network system based on specified mathematical model, there is a need to transfer the data into a format that would be useful in an analysis tool. It requires specify data format that can be easily shared between various tools or even several of transport architectures (independent form complexity). Several data sharing and exchange standards have been developed in the Intelligent Transport Systems [11]. They define a standard data format for the sharing and exchange of the transportation data mostly based on *UML* (*Unified Modeling Language*) diagrams. Other solutions, i.e. Japanese standard called *UTMS* (*Universal Traffic Management Systems*) focuses rather on the road traffic system.

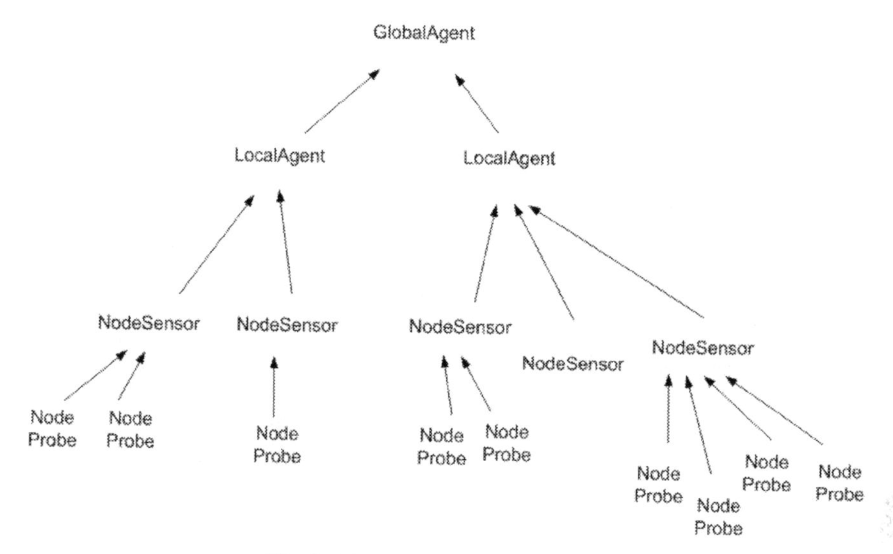

Fig. 3. Multilevel architecture schema

Still none of them is coherent with solutions proposed in this chapter, since they describe different types of network system. Moreover they are based on *UML* diagrams, which are the graphical representation of a model, but not the one, that can be simply used as an input format for any available analysis tool (computer simulator). Additionally description language for this system should be as close as real, not only to a mathematical description of the system, but to real system behavior and its parameters. In Section 4 we mentioned that the view of the network system can be realized on two levels (local and global). To do that, the tool for visualization and data processing is needed. Furthermore having this tool we can not only see the topology of the system, but also its elements and parameters. It gives us an opportunity to see the system more precisely or even make some analysis on a real data that comes from proposed multilevel agent-based architecture. Still, it requires specify data format that can be shared between tools, but since of the data exchange is done based on *UML* diagrams, there is a need used some other solution that will be more suitable. Since *UML* diagrams are mostly graphical representation of a model, we propose a solution of a *description language* for a proposed model, called *SML* (*System Modeling Language*) [16]. Format of this language is based on *XML* standards, since it is easy to use, and extendable. Moreover the format allows using the language without special tools, since *XML* is supported by many tools. Figure 4 shows a fragment of the language with appropriate elements and attributes related with mathematical model described previously. As can be seen, each element of the system is modeled as a complex element with appropriate sub-elements and attributes. The proposed language assures aggregation of dependability and functionality aspects of the examined systems. One language describes whole system and provides a universal solution for various techniques of computer analysis as an effective and suitable input for those tools. Expect easiness of potential softcomputing analysis, promising scalability and portability (between analysis tools) can be named as the main advantage of the language usage. Proposed language is easy to read and to process using popular and open-source tools; however the metadata used in this format are still a significant problem in case of file processing (large size of the file). Nevertheless since *XML* format is strongly supported by programming languages like: *Java*, *C#*, the usage as much as processing of the file can be done irrespectively from application language. As previously described (Fig. 3.) data send by *Node Probe* are combined in the *Node Sensors*. Each of these entities has assigned to it a supervisor – *Local Agent* that accumulates these files in order to create local view. This level is more compound and computational complex than previous one considering installed database and some methods that solve additional problems. In this way *XML* files transferred from the simplest level to the next one – creating views on the upper level. *Global Agent* collects this information and similarly to Local Agent combines all information included in dedicated *SML* file. As this Global Agent is the most resourceful entity it may be distributed, so it can contain more than one database. At the end, full description of the system is created, visualized and analyzed with respect to dedicated analysis tools.

```
<Node>
        <SingleCentralNode to="Wroclaw">
         <numberOfPackages>3455</numberOfPackages>
         <numberOfVehicles>8990</numberOfVehicles>
         <ManagementSystem />
         <TechnicalInfastuctureTopology numberofOrdinaryNodes="5819">
                 <timeBetweenSpecificNodes>
                     <linksBetweenNodes from="Wroclaw" to="Opole" />
                      <time>5.7</time>
                  </timeBetweenSpecificNodes>
              </TechnicalInfastuctureTopology>
          </SingleCentralNode>
       </Node>
       <Vehicle>
           <meanspeed>9.7</meanspeed>
           <capacity>678</capacity>
           <MTTF>10.05</MTTF>
           <MRT>50.98</MRT>
       </Vehicle>
```

Fig. 4. SML – fragment of the language for *DTS*

6 Critical Situations

The working point of a unified network system is defined by specific values of functional parameters (resulting from the existing infrastructure – load capacity of commodity carriers and the available number of carriers, passing transfer limits, connection quality, availability and quality of handling equipment, route selection, etc.) and reliability (mean time to elements failures, the number of repair crews, the frequency and duration of traffic jams and other problems, machine renewal time, etc.). In practice, only some elements of the system model may be treated as decision variables. For example, a system designer may adjust carrier capacities to the actual needs of the task but very often, he or she has no possibility to choose the elements base on their reliability features. For example, it is possible to choose a better throughput of the connection, but it is no chance to change the parameters of this part of the network. The appropriate operating point of the network system may be achieved thankfully to the dispatching mechanisms and the actions of organizational nature as: choosing the number of carriers and/or the number of repair crews, bypassing a blocked (overload) by traffic connections, rescheduling, etc. Dispatching decisions concerning allocation of services (functionalities) and resources can define the system reconfiguration necessary to accomplish the planned tasks.

The dependability analysis of network systems is carried out to assess the degree of risk associated with the implementation of task agreements. Note that in this case, the risk is defined and assessed as likely to ensure the system performance under certain conditions. Another important issue is the evaluation of the impact of various system parameters on defined measures of performance (performability, dependability). Dependability synthesis of network systems is based primarily on proper selection of services and resources to fulfill the functional requirements defined by users' tasks (the so-called. input tasks) – see functional – reliable models [21, 22, 27].

Optimization of system synthesis is carried out based on the minimization of potential losses resulting from breach of contract. Since the parameters and decision variables of the process of network system synthesis are determined by nominal values contained in the intervals of tolerance, though unlikely, is a scenario corresponding an operation point defined by the worst of circumstances (for example, the simultaneous maximum demand of tasks, the maximum number of long-term traffic jams, outbreaks caused by different matters, etc.). The decision variables and the parameters are very often treated as random variables within appropriate tolerance ranges. The operation point of the system may be defined together with a multidimensional solid of tolerance that is created at the appropriate confidence level.

The tolerance solid of the network system may be used as a basis for estimating the risk of system faults. It is worth noting the difference between the intended ("built-in") redundancy (functional, reliable) and pseudo-redundancies as a result of random variables distributions, and therefore both the system constructor and the dispatching mechanisms should exercise adequate caution in these situations. The set of system operation points forms a system efficient operation area defined in n-dimensional hyperspace of system parameters and decision variables. The task of synthesis of the network system can be formulated as to ensure the global task performance for a specified number of carriers, choosing the appropriate delivery route and the costs do not exceed a fixed value. Figure 5 illustrates the problem of selecting the operation point of the network system taking into account the number of carriers and repair utensils. The actual system quality is measured by the availability parameter.

The boundaries of the efficient operation area shall be determined on the basis of the acceptable costs of tasks, the maximum allowable repair time, and cost of used infrastructure. The boundaries can be set for the expected values – the hyper-planes of maximum costs of working system and the hyper-plane of the minimum, but still acceptable, system availability. It is easy to notice that the efficient operation area may consist of many operating points, which are associated with different operating costs or risk of incorrect operation of the system. It is introduced a concept of a critical operation point of the system, i.e., such an operation point within the efficient operation area that the occurrence of a single hostile incident (e.g. damage of single system element) causes a transient exit (e.g. for renewal time) beyond the area of efficiency and an additional hostile event that appears during the renewal time (e.g. a traffic jam on one of the used routes) leads to system crush (e.g. interruption of the supply chain in a just at time operating system).

A subset of the critical operation points constitutes the so-called critical efficient operation area of the system (Figure 5) corresponds to critical system operation states. The critical system state can be a simple consequence of change of "process parameters", such as raising the intensity of damage of the systems elements as a result of their use or the result of unfavorable combination of circumstances (adverse realization of random variables). For example, without necessarily changing the intensity parameter, too many carriers would be damaged at the same time, and repair crews would be overwhelmed. In extreme cases, it may lead to an avalanche of hostile events, or even to crash the system.

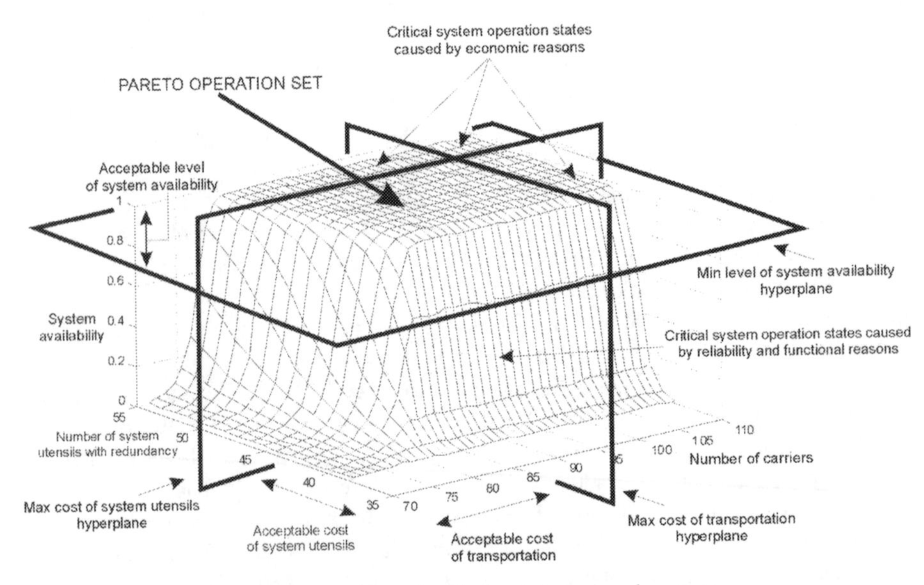

Fig. 5. General idea of critical sets for network system

7 Conclusions

We have presented a formal model of sophisticated network system including reliability, functional parameters as well as the human factor component at the necessary level of detail. The model is based on the essential elements and features extracted from the *Discrete Transport System (DTS)* and the *Computer Information System (CIS)*. We pointed the crucial conditions of the normal work of the system.

The critical situation is described and discussed to create the Pareto set – guarantying the possible safety operating points for actual network system.

The proposed approach allows performing reliability and functional analysis of the different types of network systems – for example:

- determine what will cause a "local" change in the system,
- make experiments in case of increasing volume of the commodity incoming to system,
- identify weak point of the system by comparing few its configuration,
- better understand how the system behaves.

Based on the results of simulation it is possible to create different metrics to analyze the system in case of reliability, functional and economic case. The metric could be analyzed as a function of different essential functional and reliability parameters of network services system. Also the system could be analyze in case of some critical situation (like for example a few day tie-up [24]).

The presented approach – based on two streams of data: dependability factors and the features defined by the type of business service realized – makes a starting point for practical tool for defining an organization of network systems maintenance. It is possible to operate with large and complex networks described by various – not only classic – distributions and set of parameters. The model can be used as a source to create different measures – also for the economic quality of the network systems. The presented problem is practically essential for defining and organization of network services exploitation.

References

1. Al-Kuwaiti, M., Kyriakopoulos, N., and Hussein, S.: A Comparative Analysis of Network Dependability Fault-tolerance, Reliability, Security, and Survivability. IEEE Communications Surveys & Tutorials, 11 (2), pp. 106-124 (2009)
2. Arvidsson, J.: Taxonomy of the Computer Security Incident Related Terminology. Telia CERT (2006), Retrieving date of access: May, 15, 2011 from (`http://www.terena.nl/tech/projects/cert/i-taxonomy/archive/.txt`)
3. Avizienis, A., Laprie, J.C., Randell, B.: Fundamental Concepts of Dependability. Toulouse France: LAAS-CNRS Research Report No. 1145, LAAS-CNRS (2001)
4. Avizienis, A., Laprie, J.C., Randell, B., Landwehr, C.: Basic Concepts and Taxonomy of Dependable and Secure Computing. IEEE Trans. Dependable and Secure Computing (TDSC), 1 (1), pp. 11-33 (2004)
5. Bonabeau E.: Agent-Based Modelling: Methods and Techniques for Simulating Human Systems, Proc Natl Acad Sci (2002)
6. Gao Y., Freeh V. W., Madey G. R.: Conceptual Framework for Agent-based Modelling and Simulation, Proceedings of NAACSOS Conference, Pittsburgh (2003)
7. Jennings N. R.: On Agent-Based Software Engineering, Artificial Intelligence 117, Elsevier Press, April 2000, pp. 277-296, (2000)

8. Kołowrocki, K.: Reliability of Large Systems. Amsterdam-Boston-Heidelberg-London-New York-Oxford-Paris-San Diego-San Francisco-Singapore–Sydney-Tokyo: Elsevier (2004)
9. Kyriakopoulos, N., Wilikens M.: Dependability and Complexity: Exploring Ideas for Studying Open Systems, EN. Brussels, Belgium: EC Joint Research Centre (2001)
10. Lapie J. C.: Dependability: Basic Concepts and Terminology. New-York, NY, Wien: Springer-Verlag (1992)
11. Liu H., Chu L., Recker W.: Performance Evaluation of ITS Strategies Using Microscopic Simulation, Proceedings of the 7th International IEEE Conference on Intelligent Transportation Systems, 2004, pp. 255-270 (2004)
12. Mascal C. M., North M. J.: Tutorial on Agent-Based Modelling and Simulation, Winter Simulation Conference, (2005)
13. Mazurkiewicz, J., Walkowiak, T., Nowak K.: Fuzzy Availability Analysis of Web Systems by Monte-Carlo Simulation. In Lecture Notes in Computer Science. Lecture Notes in Artificial Intelligence, pp. 616-624. Berlin, Heidelberg: Springer-Verlag (2012)
14. Melhart, B., White, S.: Issues in Defining, Analyzing, Refining, and Specifying System Dependability Requirements. In Proc. of the7th IEEE International Conference and Workshop on the Engineering of Computer Based Systems (ECBS 2000), Apr. 3-7, 2000 pp. 334-340, Edinburgh, Scotland, UK: IEEE Computer Society (2000)
15. Mellouli S., Moulin B., Mineau G. W.: Laying Down the Foundations of an Agent Modelling Methodology for Fault-Tolerant Multi-agent Systems, ESAW 2003, pp. 275-293 (2003)
16. Michalska K., Mazurkiewicz J.: Functional and Dependability Approach to Transport Services Using Modeling Language; Computational Collective Intelligence – Technologies and Applications – ICCCI 2011, 3rd International Conference, Gdynia, Poland, September 2011, Proceedings, Part II, LNAI 6923, Springer-Verlag Berlin Heidelberg 2011, P. Jędrzejowicz et al. (Eds.), pp. 180-190 (2011)
17. Michalska, K., Walkowiak, T.: Hierarchical Approach to Dependability Analysis of Information Systems by Modeling and Simulation. In Andre Cotton et al. (Eds.), Proceedings of the 2nd International Conference on Emerging Security Information, Systems and Technologies (SECURWARE 2008), Cap Esterel, France, 25-31 August, 2008, pp. 356-361, Los Alamitos: IEEE Computer Society Press (2008)
18. Michalska, K., Walkowiak, T. (2008). Modelling and Simulation for Dependability Analysis of Information Systems. In Jerzy Świątek et al. (Eds.), Information Systems Architecture and Technology. Model Based Decisions, pp. 115-125, Wroclaw: University of Technology (2008)
19. Nowak, K.: Modelling of Computer Systems – an Approach for Functional and Dependability Analysis. K. Kołowrocki, J. Soszyńska-Budny (Eds.), Journal of Polish Safety and Reliability Association, Summer Safety and Reliability Seminars (SSARS 2011), 1, pp. 153-161 (2011)
20. Walkowiak, T., Mazurkiewicz, J.: Availability of Discrete Transportation System Simulated by SSF Tool. In Proceedings of International Conference on Dependability of Computer Systems, Szklarska Poreba, Poland, June, 2008, pp. 430-437, Los Alamitos: IEEE Computer Society Press (2008)
21. Walkowiak, T., Mazurkiewicz, J.: Functional Availability Analysis of Discrete Transportation System Realized by SSF Simulator. In Proceedings of the 8th International Conference 'Computational Science – ICCS 2008', part I, Krakow, Poland, June 2008, pp. 671-678, Berlin, Heidelberg: Springer-Verlag (2008)

22. Walkowiak, T., Mazurkiewicz, J.: Algorithmic Approach to Vehicle Dispatching in Discrete Transportation Systems. In Jarosław Sugier et al. (Eds.), Technical Approach to Dependability, pp. 173-188, Wroclaw: Wroclaw University of Technology (2010)
23. Walkowiak, T., Mazurkiewicz, J.: Functional Availability Analysis of Discrete Transportation System Simulated by SSF Tool. International Journal of Critical Computer-Based Systems, 1 (1-3), pp. 255-266 (2010)
24. Walkowiak, T., Mazurkiewicz, J.: Soft Computing Approach to Discrete Transportation System Management. In Lecture Notes in Computer Science. Lecture Notes in Artificial Intelligence. vol. 6114, pp. 675-682, Berlin, Heidelberg: Springer-Verlag (2010)
25. Volfson, I.E.: Reliability Criteria and the Synthesis of Communication Networks with its Accounting. J. Computer and Systems Sciences International, 39 (6), pp. 951-967 (2000)
26. Xiaofeng, T., Changjun, J., Yaojun, H.: Applying SOA to Intelligent Transportation System. In Proceedings of the IEEE International Conference on Services Computing, Vol. 2, July, 11-15, 2005, pp. 101-104, Orlando, Florida: IEEE Computer Society (2005)
27. Zamojski, W., Caban, D.: Assessment of the Impact of Software Failures on the Reliability of a Man-Computer System. In Proc. of the Conference on European Safety and Reliability (ESREL), 2005, pp. 2087-2090, Gdynia-Sopot-Gdansk: A. A. Balkema (2005)
28. Zhou, M., Kurapati, V.: Modelling, Simulation, & Control of Flexible Manufacturing Systems: A Petri Net Approach. London, UK: World Scientific Publishing (1999)
29. Zhu, J., Zhang, L.Z.: A Sandwich Model for Business Integration in BOA (Business Oriented Architecture). In Proceedings of the IEEE Asia-Pacific Conference on Services Computing (APCSC), 2006, pp. 305-310, Washington, DC: IEEE Computer Society (2006)

Model Transformation for Multi-objective Architecture Optimisation of Dependable Systems

Zhibao Mian, Leonardo Bottaci, Yiannis Papadopoulos, Septavera Sharvia, and Nidhal Mahmud

Computer Science Department,
University of Hull, HU6 7RX, UK
{Z.Mian@2009.,L.Bottaci@,Y.I.Papadopoulos@,
s.sharvia@,N.Mahmud@}hull.ac.uk

Abstract. The promise of model-based engineering is that by use of an integrated and coherent system model both functional and non-functional requirements may be analysed, implemented and tested in a rigorous and cost-effective manner. An important part of model-based engineering is the use of analysis and design languages. The Architecture Analysis Design Language (AADL) is a new modelling language which is increasingly being used for high dependability embedded systems development. Such languages are ideally suited to model-based engineering but the use of new languages threatens to isolate existing tools which use different languages. This is a particular problem when these tools provide an important development or analysis function. System optimization is such a function.

System designers seek an optimal trade-off between high dependability and low cost. For large systems, the design space of alternatives with respect to both dependability and cost is enormous and too large to investigate manually. For this reason automation is required to produce optimal or near optimal designs.

HiP-HOPS is a mature, state of the art, dependability analysis and optimisation method and tool. HiP-HOPS requires, as input, the local failure behaviour of the system components together with the inter-component failure propagation behaviour. For optimisation, component variability information is also required.

The integration of tools such as HiP-HOPS into a model-based engineering environment requires that these tools have suitable access to the system model. Without proper integration, additional system information must be input at additional cost and risk of inconsistency.

This paper shows how model transformation may be used to integrate a multi-objective optimization method and tool into a model-based engineering environment. To illustrate the transformation method it is applied in a case study; where, drawing from the results of the optimisation, we highlight the potential value of this work for model-based design.

Keywords: MBE, dependability analysis, model transformation, ATL, AADL, HiP-HOPS, architecture optimisation.

W. Zamojski and J. Sugier (eds.), *Dependability Problems of Complex Information Systems*,
Advances in Intelligent Systems and Computing 307, DOI: 10.1007/978-3-319-08964-5_6

1 Introduction

1.1 Model-Based Engineering and System Optimisation

Model-based engineering is used to design engineering systems in which models are the central artefacts through the lifecycle of a system development process. Model-based engineering, as argued in [1], allows a systematic analyses of system architecture early and throughout the development life cycle. This can provide higher confidence that the integrated system will meet specific design goals such as performance, timing and dependability-related requirements. Furthermore, model-based engineering enables a more cost effective development and system integration process.

Recent work in this area has focused on the development of languages and notations that aim to progressively refine requirements models and design models to automatically drive the development and then verification of complex systems. These include general purpose modelling languages such as Unified Modelling Language (UML) [2] and SysML [3]. More recently, Architecture Description Languages (ADLs) such as AADL (Architecture Analysis and Design Language as described in [4]) and EAST-ADL (Electronics Architecture and Software Technology - Architecture Description Language as described in [5]) have emerged as potential future standards for model-based design of embedded systems in aerospace, automobile and avionics industries.

Beyond the modelling of "normal" behaviour, these languages also incorporate error modelling semantics which enables dependability analysis. For example, the Society for Automotive Engineer (SAE) published an Error Model Annex document [6] to complement the AADL with capabilities for dependability modelling. One of the advantages of the Error Model Annex is that it supplies a notation used for modelling the failure information on the original AADL architecture model. This kind of error annotation enables the dependability analysis to consider both intra- and inter-component error models, which is considered important for dependability analysis [7].

The design of dependable systems must address both cost and dependability concerns. For example, the cost of motor vehicles can be reduced by developing distributed flexible subsystems for functions that include steering and braking [8]. The complexity of this design space is recognised [9-10]. One problem is that a number of architectures may potentially meet the dependability requirements both technically and economically. In such architectures, any shared information and hardware resources may allow a large number of different configuration options. This greatly expands the already large design space and severely hampers the identification of the most dependable designs with minimal costs. Another problem is that there may be no solution that satisfies all the requirements. In this case, the designer must find those solutions that achieve the key requirements with the best possible trade-offs between dependability and cost. The consideration of the trade-off between objectives is an inherent part of multi-objective optimisation.

The Architecture Analysis and Design Language (AADL) has many advantages for model-based design but is relatively new and consequently there is a lack of tools that enable dependability analysis and optimization of AADL models [10]. Three challenges therefore face designers:

(1) How to ensure effective prediction of quality attributes such as dependability, via use of automated model-based analysis techniques?
(2) How best to introduce optimisation into the MBE process? If the model lacks the information required for optimisation, e.g. a scheme for representing variability, how should the model be extended to represent variability?
(3) More generally, how to use existing tools to extend the range of analyses available in a particular modelling language?

1.2 An Approach to System Optimisation for AADL Models

The approach advocated in this paper is to exploit existing dependability analysis and architecture optimisation techniques and tools. The challenge is to ensure that such tools are properly integrated into a model-based engineering process. HiP-HOPS (Hierarchically Performed Hazard Origin and Propagation Studies) [9] is a state-of-the-art model-based system dependability analysis and optimisation technique. Unfortunately, HiP-HOPS requires that the system to be optimised is expressed as a HiP-HOPS model using the HiP-HOPS modelling language. HiP-HOPS requires, as input, the local failure behaviour of the system components together with the inter-component failure propagation behavior. For optimisation, component variability information is also required. The integration of tools such as HiP-HOPS into a model-based engineering environment requires that these tools have suitable access to the system model. Without proper integration, additional system information must be input at additional cost and risk of inconsistency.

This problem can be overcome by transforming the AADL model into an equivalent HiP-HOPS model. More specifically, the AADL dependable model must be transformed into a HiP-HOPS model that captures the relevant component structure, topology and local failure information required for the HiP-HOPS analysis.

2 Background

Multi-objective optimisation problems, as argued in [9-10], should be approached systematically with the aid of optimisation techniques and computerised algorithms. An introduction to the model-based optimisation field is given by [11]. A wider survey of literature on architectural optimisation techniques is [12].

As argued in [10], to find a suitable or optimal architecture design is difficult and some automation is needed. One key issue facing system designers is how to optimise system architectures throughout the whole system development lifecycle.

Methods and tools for performing multi-objective architecture optimisation includes work based on Reliability Block Diagrams (RBDs) model [13] and, more recently, the HiP-HOPS method [8-9], [14]. HiP-HOPS is a model based dependability analysis and architecture optimisation technique. HiP-HOPS incorporates a fast algorithm for bottom up dependability analysis via automatic synthesis of fault trees and

Failure Models and Effects Analyses (FMEAs). Recently, HiP-HOPS has combined with meta-heuristics (Pareto-based Genetic Algorithms) [10] to assist in the automatic evolution of design models that can meet dependability and cost requirements. By using genetic algorithms, HiP-HOPS is able to explore the space of variations of a model and by evaluating the dependability and cost of the various model variations, HiP-HOPS is able to solve difficult multi-objective (cost and dependability) optimisation problems.

In the context of ADLs, Walker et al. [10] presents a multi-objective optimisation approach based on EAST-ADL. In this approach, three objectives, i.e., dependability, timing and cost were evaluated. The system dependability is evaluated by HiP-HOPS by transforming the EAST-ADL model into a HiP-HOPS model. EAST-ADL's variability management mechanism was used to specify the alternative implementations and thus define the design space. AADL, however, has no such scheme to define the search space of alternatives. A scheme for representing component variability is needed for optimisation of AADL models.

Related work has been done on the development of tools for multi-objective optimisation of software architectures. One tool is ArcheOpterix [15], which is based on AADL and potentially allows automatic optimisation of AADL specifications. Two quality metrics, i.e., data transmission reliability and communication overhead were evaluated. The tool was extended to enable reliability, cost and response time optimisation of AADL models shown in [16]. In this extension, only simple component redundancy allocation was used as a reliability improvement. Compared to the use of a set of component alternatives, this limits the design space.

Another tool is AQOSA (Automated Quality-driven Optimisation of Software Architecture) [17-18], for automated software architecture optimisation that allows multiple quality attributes (processor utilisation, response time, data flow latency, safety and cost). The tool is designed to use model transformation technology to convert input models (e.g. from AADL) into an intermediate model (AQOSA-IR) that can be used as the basis of the optimisation process. To generate the design space, alternative components are provided by a repository. A set of external objective function plugins provides the evaluation that drives the search process. AQOSA is designed to be independent of any given domain specific language and hence needs model transformation technology to generate analysis models from other architecture models to perform the optimisation. There is, however, no detailed work showing how the variability of alternative AADL components is represented and how the AADL dependable model can be transformed to AQOSA for AADL architecture optimisation based on dependability and cost.

In summary, there is a lack of analysis techniques and tools that can perform a dependability analysis and optimisation of AADL models. It is not always possible or best to develop and analyse systems in a single model. Different models are implemented in different languages. These include UML models, program code, interface specifications, data schemas, component descriptors and etc. [19]. Due to the use of varied models, transformations between models are necessary. The model transformation method has been used by a number of researchers [7], [20-25]. Czarnecki and Helsen [19] surveyed and analysed the domain of existing model transformation languages including model to model and model to text transformations in the literature and identified commonalities and variability among them.

3 Model Transformation Overview

There are similarities and differences in the AADL and HiP-HOPS modelling concepts. Both languages use the concepts of component, port and connection although detailed semantics differ. Error models in AADL are state machines which describe how the state of the component changes in response to events or the state of other components as observed at input ports. An omission of input or an internal component failure is an example of an event that might cause a transition to a component-failed state. In HiP-HOPS, error models for components are fault trees, i.e. local Boolean failure expressions to describe how each component may fail based on its internal malfunctions or input error deviations.

For error model transformation, we adopt the state machine to fault tree conversion algorithm shown in [24-25] rather than using the Direct Graph (DG) shown in [7]. This conversion algorithm preserves the temporal properties of the state-machine. The transformation method is similar as shown in [22]. The transformation concepts shown in [22] are used because their work is related to this research. Our transformation, however, is from a different model (AADL) and the scope is broader, aiming to encompass not only the dependability analysis but also the optimisation and temporal analysis [24] capabilities of HiP-HOPS.

In AADL, the connections between components form the main paths of error propagation through the system. In HiP-HOPS, the Line element describes how events, typically error events, propagate from one component to another. A Line element is associated with each input port. HiP-HOPS can automatically generate a system-wide fault tree from the locally defined component fault trees and the propagation information contained in the HiP-HOPS Line elements.

3.1 Translation of AADL Component Error Model to HiP-HOPS Failure Expressions

At the highest level of abstraction, the transformation consists of two parts. One part is concerned with the component specific error behaviour and the other part is concerned with inter-component error propagation. Structurally, the model transformation transforms AADL components into HiP-HOPS components and constructs HiP-HOPS Line objects from information in AADL components and connections. More specifically, for a given component, the HiP-HOPS failure expression (local fault tree) can be derived from the AADL error state machine, guard_in and guard_out expressions. The HiP-HOPS Line elements can be derived from the AADL connections.

An AADL component error model is a state machine in which component behaviour is described in terms of states and transitions between states caused by error events. Figure 1 shows an example error state machines for component `BSCU` from the case study presented later. The component is initially in the `ErrorFree` state. If the component fails then its state changes to `Failed2`. Once in this error state it propagates the event `Loss_Data` from `BSCU.Output1`. When the component loses input,

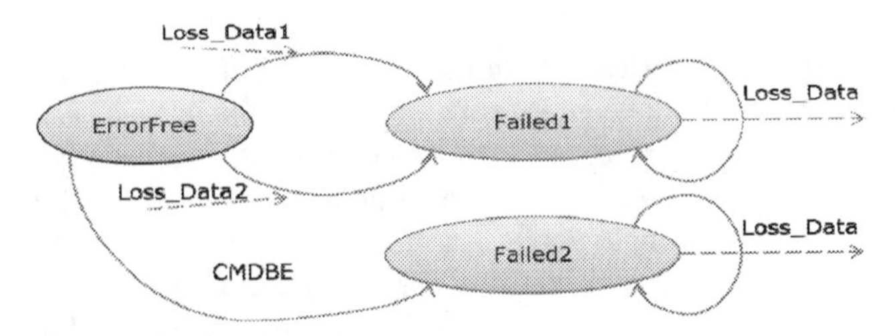

Fig. 1. Error state machine for component BSCU

either `Loss_Data1` or `Loss_Data2`, it changes to the error state `Failed1`. Once in this error state, the component propagates output error event `Loss_Data`. This event may in turn cause a transition in the error model of some other connected component.

The first stage of the transformation is to transform the component error models of each AADL component into a HiP-HOPS component fault tree expression. Each possible output of the error state machine is a top event for a fault tree. For example, in the machine shown in Figure 1, the output propagation (`Loss_Data`) is the top event and the equivalent failure expression or fault tree is:

```
Loss_Data = Loss_Data1 OR Loss_Data2 OR CMDBE                    (1)
```

This failure expression is constructed by identifying paths from the initial state to the final states corresponding to the top event. The formal algorithm is given in [23-25]. Below the top event is an OR-gate with one input for each path from the initial state to a final state corresponding to the top event. An OR-gate is used because each path represents an alternative way of reaching the final state. To traverse a path to the final state, each event that controls a state transition on that path must occur hence a path is represented by an AND-gate in which the inputs are the events that occur on the path.

3.1.1 Mapping Error States and Error Events to Component Ports

Returning to the example of Figure 1, notice the absence of port names in the expression (1) above. Error events enter and leave components via ports. The HiP-HOPS name for an error event consists of a basic event name, known also as a failure class or generic error followed by a port identifier. The notation <FailureClass>-<PortName> indicates the type of failure and the port from which it propagates.

There are six input ports for the BSCU (Figure 3). To represent port information in HiP-HOPS, expression (1) would be written as

```
Loss_Data-Output = Loss_Data1-Input1 OR … Loss_Data1-Input6
                   OR
                   Loss_Data2-Input1 OR … Loss_Data2-Input6
                   OR CMDBE                                      (2)
```

Equation (2) specifies that when the component receives an input error propagation `Loss_Data1` through any input port 1-6, the component will propagate the error `Loss_Data` through its output port `Output`. The port and event name information required to construct the HiP-HOPS, expression (2) may be obtained from the guard_in error property of the AADL model. The guard_in property associates a local error event with events from other components that may propagate along a connection to an input port. For example, the AADL guard_in property at the input port `Input1` of `BSCU` might be

```
guard_in => Loss_Data1 when Input1[Loss_Data],
            mask when others
            applies to Input1;
```

This expression means that the propagation of the error event `Loss_Data` to the input port `Input1` of `BSCU` will trigger the `Loss_Data1` error event in `BSCU`. Other error events that arrive at the input port are "masked" i.e. ignored. The port name that appears in the guard_in property can be used to translate the AADL error `Loss_Data1` into a HiP-HOPS port-name qualified failure class, i.e.

`Loss_Data1 = Loss_Data-Input1` (3)

Note that if there is no guard_in error property defined for a locally defined input error propagation, then this means that no input error name mapping is required i.e., input and output error propagations have the same name. Note also that in the situation in which a component has more than one input port and no guard_in error property is defined for a given error that may propagate to those input ports then the error may propagate through any of the input ports. To translate this situation into a HiP-HOPS failure expression, an event name of the form <FailureClass>-<PortName> is created for each port. The event which is the propagation of the error to any port can then be represented as a disjunction (OR) of port qualified names. In general, each locally defined input error propagation e that appears in a state machine is transformed into a disjunction (OR) of the names constructed by appending each of the input ports to the input error propagation:

`e = e-in1 OR e-in2 OR ... e-inN` (4)

where `in1, in2, ... e-inN` are the input ports through which e may propagate to the component.

The HiP-HOPS names of error events that propagate out of a component may be constructed in an analogous manner. In the presence of a guard_out property at a port, the error name mapping at that port can be used to create the HiP-HOPS name. In the current example, suppose that there is the following guard_out property at the output port `Output1` of `BSCU`

```
guard_out => Loss_Data when self [Failed2],
             mask when others
             applies to Output1;
```

which means that the component will propagate an output error propagation called `Loss_Data` through output port `Output1` when it is in the state `Failed2`. The other error propagations propagate through this output port are "masked" i.e. not propagated out. Again, the fact that this guard_out property is associated with the port `Output1` may be used to qualify the `Loss_Data` event, i.e.

```
Loss_Data = Loss_Data-Output1                    (5)
```

The guard_out also allows the mapping of the component `Failed2` state to the output propagation `Loss_Data` (HiP-HOPS failure class `Loss_Data-Output1`), i.e.

```
Loss_Data-Output1 = Failed2 = CMDBE              (6)
```

In the absence of any guard_out error property, the output error propagations defined for a component will propagate through each output port of that component. For a given set of errors that propagate out of a component, a HiP-HOPS failure class is created for each error. For a given set of output ports, each port is used to qualify the failure class. More formally, suppose that there are number n of output ports (n >= 1), i.e., `out1`, `out2, . . . , outn`, then we obtain:

```
OutputErr = OutputErr-Out1 = OutputErr-Out2 = ...
          = OutputErr-Outn                       (7)
```

For component `BSCU`, there are three output ports: `Output, Output1 and Output2`. Thus, in the absence of any guard_out error property, based on the Boolean logic shown in (7), we obtain:

```
Loss_Data = Loss_Data-Output = Loss_Data-Output1
          = Loss_Data-Output2                    (8)
```

For component BSCU, from Boolean logic (5) and (6) we now obtain:

```
Loss_Data-Output1 = CMDBE                        (9)
```

This is the HiP-HOPS Boolean failure expression (fault tree) for the component `BSCU` and all event names except local failure event e.g. CMDBE, are expressed as <failure class>-<port name>.

3.2 Transformation of AADL connections to HiP-HOPS Lines

Using the AADL state machine to HiP-HOPS fault tree transformation described in the previous section, we can obtain a local fault tree for each component in the system. To create a whole system fault tree, HiP-HOPS needs information about how errors propagate between components. This information is represented using HiP-HOPS Lines. The HiP-HOPS Line element describes how events, typically error events, propagate from one component to another. The HiP-HOPS Line concept describes a set of connected ports. The Line contains a set of HiP-HOPS Connection objects. Each Connection describes the propagation of event to a specific port from other ports. A Line connecting two ports will have two Connections if events flow in both directions.

The information required to create HiP-HOPS Lines can be obtained from the AADL connection objects. To give a simple description of the transformation from AADL connections to HiP-HOPS Lines, consider a simple case in which only one AADL connection (called `DataConnection1`) is defined between two components `Power` and `BSCU` (Figure 3). Assume also that there is an ‘in out’ error propagation called Loss_Data which is defined in an error model and this error model is associated with both components Power and BSCU. The partial AADL description of this connection is

```
DataConnection1: data port Power.Output -> BSCU.Input1;
```

For this AADL connection, the error events will propagate from the source port Power.Output to the destination port BSCU.Input1. In particular we can associate a connection logic failure expression (called HiP-HOPS PortExpression) with port BSCU.Input1 which describes the failure at component BSCU in terms of the output failure at Power. Thus the HiP-HOPS Line for the connection from Power.Output to BSCU.Input1 would be constructed as follows

```
<Line> <Type>Directed</Type>
  <Connections>
    <Connection>
      <FailureClass> // failure in component
         Loss_Data
      </FailureClass>
      <Port>BSCU.Input1</Port> // propagated into port
      <PortExpression> // failure propagated when
         Loss_Data-Power.Output
      </PortExpression>
    </Connection>
  </Connections>
</Line>
```

Each Line element contains a list of Connections. Each Connection describes how errors at one or more output ports (e.g. Loss_Data at Power.Output) propagate to an error at an input port (e.g. Loss_Data at BSCU.Input1). The <Port> attribute identifies the port to which the error propagates. Since the Line is directed the error will propagate from the port (Power.Output) to the port (BSCU.Input1). The PortExpression element is a Boolean expression containing the names of other ports on the Line. The PortExpression describes the events at other components, i.e. Loss_Data from Power.Output, which causes an event, in this case, Loss_Data at the Input1 port of component BSCU.

The transformation of the above example AADL connection to HiP-HOPS Line is relatively straightforward as errors to the port BSCU.Input1 can come only from one port, i.e. Power.Output. The transformation transforms the AADL connection's destination port to HiP-HOPS Connection destination port and the AADL connection's source port to HiP-HOPS PortExpression. The AADL output error propagation Loss_Data is transformed to a HiP-HOPS FailureClass and the portExpression is constructed in the style of < FailureClass >-<portname>.

Whereas an AADL connection joins only two ports, a HiP-HOPS Line may connect any number of ports. For each port in a Line to which error events may propagate, there is a HiP-HOPS Connection object that specifies how error events may propagate to that port from other ports in the Line. The HiP-HOPS Line also maps the names of events at source components to names of events in the destination component. This is an important difference between AADL connections and HiP-HOPS Lines. HiP-HOPS models multiple AADL connections that fan-in to a single destination port using one HiP-HOPS Line. Since an error event may originate from any of the fan-in components, the transformation introduces the OR logic operator into the port expression.

To generalise, the corresponding algorithm for transforming AADL connections to HiP-HOPS Lines is given in Figure 2. In a hierarchically structured model, the algorithm is applied to the top-level system and any sub-system.

```
Let DestPorts = {c : sys.Connections • c.destination};
ConnsSameDest =
  {p: DestPorts • {c: sys.Connections | c.destination = p}}
Lines = {cd : ConnsSameDest • Line({c : cd •
   {e : c.source.component.errorsPropagated •
      ConnectionH(c.destination,
                     OR{c : cd | e ∈c.source.component.errorsPropagated
                                 • e-c.source.name})}})}
```

Fig. 2. The formal algorithm for transforming AADL Connections into HiP-HOPS Lines

The formal description of the algorithm shown in Figure 2 should be read as follows. The variable `DestPorts` is defined to be the set generated (• denotes generator operator) by collecting the destination port of each connection in the system. The system is represented by the variable `sys`. The variable `ConnsSameDest` is defined to be a set of connection sets. In each connection set, all the connections share the same destination port. For each destination port, a set of connections to that port is generated by filtering (| denotes filter operator) the connections with a destination equal to a given destination port. The variable `Lines` is the set of HiP-HOPS Line objects. For each set of connections in `ConnsSameDest`, a HiP-HOPS Line object is constructed. A Line is constructed from a set of HiP-HOPS Connection (`ConnectionH`). A HiP-HOPS Connection is constructed for each failure class that is propagated from any component that is at the source of any connection in the set of connections to a given destination port. The HiP-HOPS PortExpression is a disjunction because the error may propagate from any of the source components, hence

`OR{c : cd • e-c.source.name}` where `e` is a failure class and `c.source.name` is a port name. The operator OR denotes e-c1.source.name OR e-c2.source.name .. e-cn.source.name, for each connection ci in cd.

In this definition, `c.source.component.errorsPropagated` denotes the set of output error propagations (HiP-HOPS failure classes) from the component at the source of connection `c`. These error propagations can be obtained from the error model of the source component. Each failure class collected, denoted `e`, is used to qualify the connection source port name, i.e. `e-c.source.name`.

HiP-HOPS allows a number of abbreviated syntax forms in order to improve readability. If a Connection lacks a failure class then the PortExpression applies to all failure classes. If a model is not intended to be human-readable then such abbreviations are unnecessary. Omitting such abbreviation typically simplifies the model to model transformation and is the approach adopted in this work.

One challenge for the optimisation of AADL models, is how to represent model 'variability' in the AADL system model. Variability includes the possibility of designating one or more alternatives to a given component or subsystem. Variability is a prerequisite for optimisation, because it creates the design space of alternative designs which needs to be explored in order to seek the best solutions.

Clearly, for automated optimisation, the variability must be constrained. In the optimisation method considered in this paper, any component or subsystems may be associated with one or more alternatives. Each alternative component has an equivalent function but a different dependability and cost. The optimisation process searches the large space of possible designs defined by the combinations of possible choices, and uses optimisation heuristics such as genetic algorithms to obtain optimal or near optimum designs.

In AADL, however, there is no direct means of modelling component alternatives and other optimisation parameters. Mian et al. [26] introduces a method which allows the AADL designer to specify variable elements of the system model. That method is used in this paper. To enable optimization, additional information on component alternatives is required. These alternatives provide options in terms of trade-off between dependability and cost. Each component is annotated with its alternatives, i.e. components performing identical functions but with differing costs and failure rates.

3.3 Model Transformation Implementation

The Eclipse Modelling Framework (EMF) [27] is a modelling framework and a highly flexible tool platform. Different plugins for different models can be developed based on EMF. The Open-Source AADL Tool Environment (OSATE) developed by SEI [28] is a set of plug-ins based on Eclipse and the EMF. The OSATE plug-ins were used in the work reported here and the model transformation method has been implemented as an OSATE plugin.

Model to model transformation languages should be well suited for our semantic mapping transformation, since both input and output are models. We chose the ATLAS Transformation Language (ATL) [29-30] which is a hybrid language containing a mixture of declarative and imperative constructs. The declarative rule-base language hides much of the complexity of navigating the AADL source model. In addition, ATL has been shown to be effective for similar model transformations [22].

4 Case Study

4.1 System Description

The aircraft wheel brake system model is adapted from the Aerospace Recommended Practice [31] aircraft wheel brake system, which is also presented in [32]. Figure 3 shows the basic system structure and Figure 4 shows the corresponding AADL description of the wheel brake system. The primary function of the wheel brake system is to provide safe braking function for aircraft which requires supplying correct pressure and preventing skidding. Braking can be either manual or automatic. Manual braking is controlled via brake pedals, while automated braking does not require pedal application. The automated braking is realized via Autobrake function which allows the pilot to provide the deceleration rate prior to takeoff or landing.

The braking system operates in one of two modes, Normal or Alternate. In Normal braking mode, GreenPump provides the required hydraulic pressure, and Alternate mode is held on standby. If failure occurs on Normal mode, the system moves to Alternate mode and hydraulic power is generated by the BluePump. In the original ARP 4761 example, another backup mechanism is in place lest both of the pumps fail. In this paper, however, it has been deliberately excluded to demonstrate how HiP-HOPS can be used to help guide the analysis process and the identification of potential safety measures.

The Brake System Control Unit (BSCU), is the digital controller which accepts inputs to compute braking and anti-skid commands. In its Normal operational mode, BSCU receives information from various input sources. It obtains brake pedal positions as input and processes this information to produce command signals to the brakes. When Autobrake is true, deceleration rate and aircraft speed are used to calculate the brake command. BSCU also monitors signals which indicate certain critical aircraft and system states to provide correct brake function, generate warnings, indications and maintenance information to other system.

Two hydraulic pressure lines are used: the Green line, powered by the GreenPump (Normal) and the Blue line, powered by the BluePump (Alternate). The GreenValve and the BlueValve are used to control the pressure from the GreenPump and BluePump respectively. The SelectorValve is located across the Green and Blue hydraulic lines, and selects only one of the two hydraulic systems to provide pressure to the brakes. This pressure is relayed to the corresponding meter valves, CMD/ASMeterValveG and CMD/ASMeterValveB respectively. The meter valves take two inputs: the incoming pressure and the valve position command. The valve position is adjusted to output the required amount of pressure based on the command from the BSCU.

The system switches to Alternate mode when the pressure along the green line falls below a threshold. Once BSCU identifies that Alternate line should be activated, it sends an OnAlternate signal which commands SelectorValve to switch to the Blue line. Once the system switches to Alternate, it will not revert back to Normal. The component labelled WBS is the pressure output block, the components NormalP and AlternateP serve only to propagate failures.

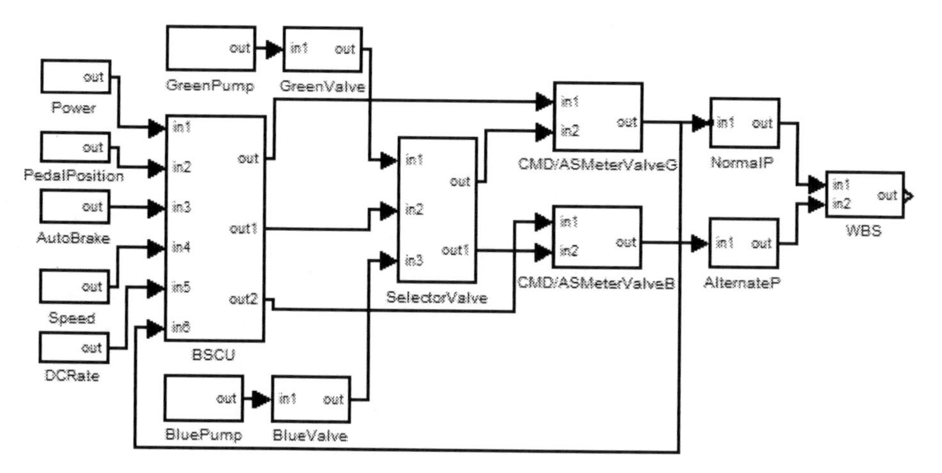

Fig. 3. The basic system structure of aircraft wheel brake system

```
system implementation AWBS.SystemImpl1
  subcomponents
    Power: device DeviceType1.Power;
    PedalPosition: device DeviceType1.PedalPosition;
    AutoBrake: device DeviceType1.AutoBrake;
    NormalP: device DeviceType2.NormalP;
    AlternateP: device DeviceType2.AlternateP;
    CMDASMeterValveG: device DeviceType3.CMDASMeterValveG;
    . . .
    --Other components omitted
    . . .
    WBS: device DeviceType3.WBS;
    SelectorValve: device DeviceType4.SelectorValve;
    BSCU: device DeviceType5.BSCU;
  connections
    DataConnection1: data port Power.Output -> BSCU.Input1;
    DataConnection2: data port PedalPosition.Output -> BSCU.Input2;
    DataConnection3: data port AutoBrake.Output -> BSCU.Input3;
    . . .
    --Other connections omitted
    . . .
    DataConnection17: data port CMDASMeterValveG.Output -> BSCU.Input6;
    DataConnection18: data port NormalP.Output -> WBS.Input1;
    DataConnection19: data port AlternateP.Output -> WBS.Input2;
end AWBS.SystemImpl1;
```

Fig. 4. The AADL description for the aircraft wheel brake system

4.2 Failure Data

The AADL Error Model Annex is used to model the system failure behavior. For simplicity, each component is assumed to be vulnerable to one internal failure which leads to the omission of component output. Other types of component failure

```
package AWBS_ErrorModel
public
  annex Error_Model {**
    error model Basic --basic error model
    features
      ErrorFree: initial error state;
      GreenPumpBE, GreenValveBE,BluePumpBE, BlueValveBE, GCMDASBE,
      BCMDASBE, SelValveBE, CMDBE, PowerBE, PedalPositionBE, AutoBrakeBE,
      SpeedBE, DCRateBE: error event;
      Failed, Failed1, Failed2, Failed3: error state;
      Loss_Data: in out error propagation {occurrence => fixed 0.8};
      Loss_Data1, Loss_Data2: in error propagation;
    end Basic;

    error model implementation Basic.Power
    --error transitions omitted
    end Basic.Power;

    error model implementation Basic.GreenValve
    --error transitions omitted
    end Basic.GreenValve;

    error model implementation Basic.BSCU
    transitions
      ErrorFree -[ Loss_Data1 ]-> Failed1;
      ErrorFree -[ Loss_Data2 ]-> Failed1;
      ErrorFree -[ CMDBE ]-> Failed2;
      Failed1 -[ out Loss_Data ]-> Failed1;
      Failed2 -[ out Loss_Data ]-> Failed2;
    properties
      occurrence => poisson 1.0e-10 applies to CMDBE;
    end Basic.BSCU;
    …
  **};
end AWBS_ErrorModel;
```

Fig. 5. AADL error model type definition and error model implementation for component Power, GreenValve and BSCU

(for example, commission or value failure) are not discussed but may be treated analogously. The internal failures for components GreenPump, GreenValve, BluePump, BlueValve, CMD/ASMeterValveG, CMD/ASMeterValveB and SelectorValve are denoted as GreenPumpBE, GreenValveBE, BluePumpBE, BlueValveBE, GCMDASBE, BCMDASBE, SelValveBE respectively. Internal failure in the BSCU (the command unit) is denoted as CMDBE. The input to the BSCU comes from Power, PedalPosition, AutoBrake, Speed and DCRate component. Internal failures in these input components are denoted as PowerBE, PedalPositionBE, AutoBrakeBE, SpeedBE, and DCRateBE respectively.

Figure 5 shows the AADL error model type definition and error model implementation for component Power, GreenValve and BSCU. The error state machine shown in Figure 1 is specified in the error model implementation of Basic.BSCU shown in

```
device DeviceType5
  features
    Input1: in data port; Input2: in data port; Input3: in data port;
    Input4: in data port; Input5: in data port; Input6: in data port;
    Output: out data port; Output1: out data port; Output2: out data port;
  properties
    Optimisation_Attributes::Optimise => true;
end DeviceType5;

device implementation DeviceType5.BSCU
  properties
    Optimisation_Attributes::Exclude_From_Optimisation => false;
    Optimisation_Attributes::Cost => "50";
    Alternatives::List_of_Alternatives => (device DeviceType5.BSCU2,
                  device DeviceType5.BSCU3, device DeviceType5.BSCU4);
  annex Error_Model {**
    model => AWBS_ErrorModel::Basic.BSCU;
    guard_in =>
      Loss_Data1 when Input1[Loss_Data],
      mask when others
      applies to Input1;
    guard_in =>
      Loss_Data2 when Input2[Loss_Data] AND
      (Input3[Loss_Data] OR Input4[Loss_Data] Or Input5[Loss_Data]),
      mask when others
      applies to Input2,Input3,Input4,Input5;
    guard_out =>
      Loss_Data when self[Failed2],
      mask when others
      applies to Output1;
  **};
end DeviceType5.BSCU;


device implementation DeviceType5.BSCU2
  properties
    Optimisation_Attributes::Exclude_From_Optimisation => true;
    Optimisation_Attributes::Cost => "20";
  annex Error_Model {**
    model => AWBS_ErrorModel::Basic.BSCU2;
  **};
end DeviceType5.BSCU2;

--Implementation of DeviceType5.BSCU3 and DeviceType5.BSCU4 similar to above
but cost and error model differ
```

Fig. 6. Associated AADL error model, guard_in and guard_out error model properties for component BSCU and alternative implementations of this component

Figure 5. Figure 6 shows the association of error model implementation Basic.BSCU to component BSCU, the guard_in and guard_out error properties for component implementation DeviceType5.BSCU and the alternative implementations of this component. Note the guard_in and guard_out error properties shown in `Error_Model` in DeviceType5.BSCU. These error properties specify conditions under which the input or output error propagations occur. From Figure 6, the Optimise property with value true means that alternatives for device DeviceType5 should be considered in the optimisation process. This property allows the optimisation process to be applied selectively to parts of the system. For the implementation component, DeviceType5.BSCU2, the property Exclude_From_Optimisation property with value true means that this alternative component should not be used as a replacement, i.e., is not an available alternative. This allows the designer fine-grained control over the alternatives used in the optimisation. The Cost property specifies the cost of this implementation. The device DeviceType5 has three alternative implementations (DeviceType5.BSCU2, DeviceType5.BSCU3, and DeviceType5.BSCU4), which is specified in the properties via setting the property value of List_of_Alternatives. Due to space limitation we only show two alternative implementations. For each of the implementations, the cost and other optimisation properties are specified.

4.3 Analysis of the Wheel Brake System Model

HiP-HOPS produces, FTA and FMEA. The minimal cut sets show the potential hazardous combinations of component failures which lead to O-WBS.out. The analysis of the results of FTA and FMEA shows, for e.g., that the omission of Power, BSCU command unit and SelectorValve directly leads to omission of pressure on the wheel brake. The other single point failures are also identified. For a small design model, manual analysis may be manageable. But for larger system, where this architecture may be nested within a more complex design, manual analysis becomes laborious and error-prone.

4.4 Design Optimization

In this case study, we assumed that each component has four different alternatives (each with different failure rate and different cost). Components with lower failure rates have a higher cost. Table 1 summarizes the failure rates and costs data for the component alternatives. The failure rates and costs of CMD/ASMeterValves follow those of green valves and blue valves. It should be noted that the values of failure rates are not based on any empirical data, but chosen hypothetically to illustrate the method.

Table 1. The failure rates and costs data for the component alternatives

Component	Failure Rate λ	Cost
BSCU1/ SelectorValve1	1e-10	50
BSCU2/ SelectorValve2	2e-10	20
BSCU3/ SelectorValve3	3e-10	10
BSCU4/ SelectorValve4	5e-10	5
GreenPump1/Valve1/BluePump1/Valve1	1e-8	16
GreenPump2/Valve2/BluePump2/Valve2	2e-8	8
GreenPump3/Valve3/BluePump3/Valve3	3e-8	4
GreenPump4/Valve4/BluePump4/Valve4	4e-8	2

In general, there are 4 potential component architectures for each of these 8 components. The size of the design space to be explored is therefore 4^8 possible configurations. As such it is very difficult to do this kind of optimisation manually.

4.5 Optimization Results

Based on these parameters the multi-objective optimization problem is to minimize both system risk and cost.

Figure 7 shows the optimal architectures on the Pareto front. These solutions are less risky than all other more costly solutions. To obtain specific solutions from the Pareto front, the goal of the optimization was defined as:

$$\text{Risk} \leq 0.000015,\ \text{Cost} \leq 120$$

Three solutions which satisfy this constraint are presented in Table 2. The configuration shows the combination of component alternatives selected for the solutions.

The various design solutions shows different potential configurations of components to achieve the pre-defined risk and cost restrictions. BSCU and SelectorValve are highly critical components and therefore should be robust. This is illustrated by how Solution 3, which has the lowest risk among the three selected sample solutions within the restricted cost, employs the BSCU1 and SelectorValve2. The results presented here represent a preliminary step in the overall safety assessment process. The multi-objective assessment routine can be performed iteratively by adjusting design parameters (risk and cost) until requirements are met in the process of an evolving design. The optimization is automated and therefore can be repeated efficiently in the course of design iterations.

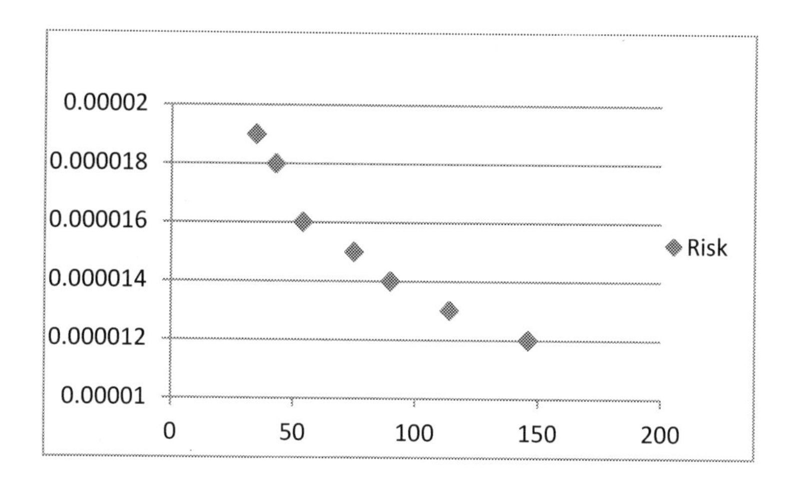

Fig. 7. The Pareto front optimal solutions

Table 2. The three solutions which satisfy the constraint: Risk ≤ 0.000015, Cost ≤ 120

Component	Solution 1	Solution 2	Solution 3
BSCU	BSCU2	BSCU2	BSCU1
BluePump	BluePump2	BluePump2	BluePump2
BlueValve	BlueValve3	BlueValve1	BlueValve3
CMD/ASMeterValveB	CMDASMe-terValveB1	CMDASMe-terValveB4	CMDASMeter-ValveB2
CMD/ASMeterValveG	CMDASMe-terValveG3	CMDASMe-terValveG1	CMDASMeter-ValveG3
GreenPump	GreenPump3	GreenPump3	GreenPump1
GreenValve	GreenValve2	GreenValve3	GreenValve3
SelectorValve	SelectorValve3	SelectorValve2	SelectorValve2
Cost	74	90	114
Risk	0.000015	0.000014	0.000013

5 Conclusion and Future Work

A model transformation method has been devised and implemented for the dependability and cost optimisation of AADL models. The direct benefit of the transformation presented in this paper is that it opens a path that will enable the AADL language to take advantage of an existing dependability analysis and optimisation technique. The technique may be used early in the design and makes the analysis of complex dependable systems practical and cost-effective. Model transformation plays a key role in model driven system development and analysis. It allows the wide application and reuse of tools.

If AADL models could be transformed into the models used by other methods then it would extend the range of analysis that could be done on AADL models. We believe that model transformation is a fundamental technique to maximise the utility of MBE because it provides a route for the exploitation of mature and tested tools in a MBE context.

Future work will consider new techniques for describing model variability. In addition to replacing a single component with an alternative component, a designer may wish to introduce other replacement patterns. For example, a replacement pattern may require that two matching components are always replaced as a pair.

References

1. Feiler, P., Gluch, D.: Model-Based Engineering with AADL-An Introduction to the SAE Architecture Analysis & Design Language. Pearson Education, USA (2012)
2. OMG: Introduction To OMG's Unified Modelling Language (UML), `http://www.omg.org/gettingstarted/what_is_uml.htm`
3. OMG: OMG Systems Modeling Language (OMG SysML version 1.3), `http://www.omg.org/spec/SysML/1.3/`

4. SAE-AS5506: Architecture Analysis and Design Language (AADL). Society of Automotive Engineers (SAE) (2006)
5. MAENAD project: EAST-ADL Domain Model Specification version V2.1.11, `http://east-adl.info/Specification/V2.1.11/EAST-ADL-Specification_V2.1.11.pdf`
6. SAE-AS5506/1: Architecture Analysis and Design Language Annex Volume 1, Annex E: Error Model Annex. Society of Automotive Engineers (SAE) (2006)
7. Joshi, A., Vestal, S., Binns, P.: Automatic Generation of Static Fault Trees from AADL Models. In: DSN Workshop on Architecting Dependable Systems, DSN 2007-WADS, Edinburgh, Scotland, UK (2007)
8. Papadopoulos, Y., Grante, C.: Evolving car designs using model-based automated safety analysis and optimisation techniques. The Journal of Systems and Software 76(1), 77–89 (2005)
9. Adachi, M., Papadopoulos, Y., Sharvia, S., Parker, D., Tohdo, T.: An approach to optimization of fault tolerant architectures using HiP-HOPS. Software Practice and Experience 41(11), 1303–1327 (2011)
10. Walker, M., Reiser, M.O., Tucci-Piergiovanni, S., Papadopoulos, Y., Lönn, H., Mraidha, C., Parker, D., Chen, D.J., Servat, D.: Automatic optimisation of system architectures using EAST-ADL. Journal of Systems and Software 86(10), 2467–2487 (2013)
11. Grunske, L., Lindsay, P., Bondarev, E., Papadopoulos, Y., Parker, D.: An outline of an architecture-based method for optimizing dependability attributes of software-intensive systems. In: de Lemos, R., Gacek, C., Romanovsky, A. (eds.) Architecting Dependable Systems IV. LNCS, vol. 4615, pp. 188–209. Springer, Heidelberg (2007)
12. Aleti, A., Buhnova, B., Grunske, L., Koziolek, A., Meedeniya, I.: Software architecture optimization methods: a systematic literature review. IEEE Transactions on Software Engineering (99) (September 2012) ISSN: 0098-5589
13. Konak, A., Coit, D.W., Smith, A.E.: Multi-objective optimization using genetic algorithms. Reliability Engineering & System Safety 91(9), 992–1007 (2006)
14. Hamann, R., Uhlig, A., Papadopoulos, Y., Rüde, E., Grätz, U., Walker, M., et al.: Semi Automatic Failure Analysis Based on Simulation Models. In: The ASME 27th International Conference on Offshore Mechanics and Arctic Engineering, OMAE 2008, Estoril (2008)
15. Aleti, A., Bjornander, S., Grunske, L., Meedeniya, I.: ArcheOpterix: An extendable tool for architecture optimization of AADL models. In: Proceedings of the 2009 ICSE Workshop on Model-Based Methodologies for Pervasive and Embedded Software, pp. 61–71 (2009)
16. Meedeniya, I., Aleti, A., Bühnova, B.: Redundancy allocation in automotive systems using multi-objective optimisation. In: Symposium of Avionics/Automotive Systems Engineering (SAASE 2009), San Diego (2009)
17. Li, R., Etemaadi, R., Emmerich, M.T.M., Chaudron, M.R.V.: Automated Design of Software Architectures for Embedded Systems using Evolutionary Multiobjective Optimization. In: Proc. of the VII ALIO/EURO (2011)
18. Etemaadi, R., Chaudron, M.R.V.: A model-based tool for automated quality driven design of system architectures. In: Proceedings of the 8th European Conference on Modelling Foundations and Applications (ECMFA 2012), Lyngby, Denmark (2012)
19. Czarnecki, K., Helsen, S.: Classification of Model Transformation Approaches. In: OOPSLA 2003 Workshop on Generative Techniques in the Context of MDA, Anaheim, USA (2006)

20. Rugina, A.E.: Dependability modelling and evaluation - From AADL to stochastic Petri nets. PhD dissertation, LAAS/CNRS (2007)
21. Rugina, A.E., Kanoun, K., Kaâniche, M.: An Architecture-based Dependability Modelling Framework Using AADL. In: 10th IASTED International Conference on Software Engineering and Applications (SEA 2006), Dallas (USA), pp. 222–227 (2007)
22. Biehl, M., Chen, D., Torngren, M.: Integrating Safety Analysis into the Model-based Development Toolchain of Automotive Embedded System. In: LCTES 2010, Stockholm, Sweden (2010)
23. Rauzy, A.: Mode automata and their compilation into fault trees. Rel. Eng. & Sys. Safety (RESS) 78(1), 1–12 (2002)
24. Mahmud, N., Papadopoulos, Y., Walker, M.: A translation of State Machines to temporal fault trees. In: International Conference on Dependable Systems and Networks Workshops (DSN-W), Chicago, USA, pp. 45–51 (2010)
25. Mahmud, N., Walker, M., Papadopoulos, Y.: Compositional synthesis of Temporal Fault Trees from State Machines. Special Issue on Modeling Dynamic Behaviors of Complex Distrib. Syst. 39, 79–88 (2012)
26. Mian, Z., Bottaci, L.: Multi-objective Architecture Optimisation Modelling for Dependable Systems. In: the 4th IFAC Workshop on Dependable Control of Discrete Systems (DCDS 2013), York University, UK (2013)
27. Steinberg, D., Budinsky, F., Paternostro, M., Merks, E.: EMF: Eclipse Modeling framework. Pearson Education, Boston (2009)
28. Feiler, P., Gluch, D., Hudak, J.: The Architecture Analysis & Design Language (AADL): An Introduction, http://www.sei.cmu.edu/reports/06tn011.pdf
29. Jouault, F., Allilaire, F., Bezivin, J., Kurtev, I.: ATL: A model transformation tool. Science of Computer Programming (72), 31–39 (2008)
30. ATLAS group: ATL: Atlas Transformation Language. ATL Starter's Guide
31. ARP 4761: Aerospace recommended practice: guidelines and methods for conducting the afety assessment process on civil airborne systems and equipment. Society of Automotive Engineering. Warrendale, PA, Tech. Rep. (1996)
32. Joshi, A., Heimdahl, M.P.E., Miller, S., Wallen, M.: Model-Based Safety Analysis. University of Minnesota Advanced Technology Center (2006)

Optimization in CIS Systems

Czeslaw Smutnicki

Wroclaw University of Technology, Wroclaw, Poland

Abstract. The chapter provides an approach for solving optimization task followed from relocation/reconstruction of distributed services in the SaaS cloud computing model in case of its malfunction, by using multicriteria evaluation with supporting simulation of possible choreographies. Beside critical survey of methods, approaches and trends observed in modern optimization, focusing on nature-inspired techniques recommended for particularly hard discrete multicriteria problems, we discuss subject of network architecture and its suscebility on attacks and malfunctions in terms of system dependability. Applicability of the methods, depending the class of stated optimization task and classes of goal function, have been also discussed.

1 Introduction

The notion Complex Information System (CIS) appearing in the literature has a broad spectrum of meaning. We understand here CIS as special class of computer system composed of workstations (clients), servers of contents or services and the net linking all these players. Generally, it works in the mode question-and-answer, although some deputed task directed to a server can be splitted and sub-ordered to next servers. One can consider CIS as a localization of various resources (hardware, services, software, databases, contents, etc.) dispersed among nodes in the net, called sometimes SaaS cloud computing model with centralised (balanced) management or web-based service, [16]. In our opinion, the last name characterize the best esential features of the system.

It is clear that events in this system have discrete character, with high dynamics of changes, unpredictive (random) set of coming events and huge size of dimension. Moreover, several procedures running in the system may be defined only as a sequence of commands or activities.

From modeling point of view CIS can be perceived as the non-stationary mixed open/closed queuening system with queues of limited length, each of which has set its own service policy. Such complex system cannot be analysed analytically, because of insuffcient power of theoretical methods. Then, many researchers consider the simulation as the most proper tool for describing and analysing behaviour of CIS.

Beside mentioned modeling aspects, a lot of attacks and malfunctions have been observed in the net, influencing on availability/unavailability of resources of some kind, thus on system dependability. In order to save the viability of service quality after the malfunction, reconfiguration of the system architecture has been

W. Zamojski and J. Sugier (eds.), *Dependability Problems of Complex Information Systems*, Advances in Intelligent Systems and Computing 307, DOI: 10.1007/978-3-319-08964-5_7

recommended. Usually several scenarios of the changes (called sometimes in the literature choreography) are possible and can be considered. Each scenario has been evaluated from several points of view. Our aim is to select the best target scenario. This leads to the task of discrete optimization (because of finished set of possible scenarios) and multicriteria (because of the number of evaluation criteria taken into account) with unusual technology of goal function evaluation (simulation). Such optimization tasks have not been considered in optimization theory so far, then the best practices from several approaches and fields are especially welcome.

We provide, beside critical survey of methods, approaches and trends observed in modern optimization, focusing on nature-inspired techniques recommended for particularly hard discrete multicriteria problems, some proposals for using these tools to solve the problem of optimal reconfiguring of CIS.

2 Optimization Technologies

Approaches employed to solve optimization goals generated by problems of on-line decision making, load balancing, task scheduling, control, planning, designing and management, significantly evolved in recent years. Cases with unimodal, convex, differentiable scalar goal functions disappeared from research labs, because a lot of satisfactory efficient methods were already developed. On the battlefield still remain very hard cases: multimodal, multi-criteria, non-differentiable, NP-hard, discrete, with huge dimensionality, with exponential increase of the number of local extremes, without apriori information about data, with random data, etc. These practical goals, generated by computer systems and networks, industry and market, evoke serious troubles observed in the process of seeking global optimum. Great effort has been done by scientist in recent years in order to reinforce power of solution methods and to fulfill expectations of practitioners. The moderate success in algorithms development strike practitioners fancy, so there is still needs for further research in this area.

3 Optimization Troubles

In the next few sections we refer to the following form of single-criterion optimization case: find $x^* \in \mathcal{X}$ so that

$$K^* \stackrel{\text{def}}{=} K(x^*) = \min_{x \in \mathcal{X}} K(x) \tag{1}$$

where x, x^*, $\mathcal{X}$ and $K(x)$ are solution, optimal solution, set of feasible solutions and scalar goal function, respectively. The form of x, $\mathcal{X}$ and $K(x)$ depends on the type of optimization task. We focus chiefly on practical discrete NP-hard problems, where $\mathcal{X}$ is discrete, $K(x)$ is nonlinear and non-differentiable. Other optimization cases (multicriteria) will be discussed in detail thereinafter.

Up to now, there has been recognized a few reasons considered as responsible for failures of the solution methods, like as: slow convergence of an optimization method to optimal or good solution, premature convergence to poor solutions, and/or high calculation cost. Main causes and effects are discussed below, illustrated by some common literature benchmarks, collected for example in [14], derived originally from Griewank, Langermann, Shakel, Rosenbrock, and others.

Multiple Extremes. This phenomenon is well illustrated by many benchmark functions, see summary in [14], starting from *six-hump camel back* function having only six local extremes located in the flat canion. The more stressful is *Griewank's benchmark* function which exhibits huge number of local extremes in every small part of the space. Hopefully, for the Griewank's function extremes are quite regularly distributed in the space. The interpretation of the latter function changes with the scale of the view; the general overview suggests classical convex function, medium-scale view suggests existence of a few local extremes, and whereas high-scale zoom indicates complex structure of numerous local extremes. Theoretically, due to regularity of the surface, one can easily define strategic search directions which lead quickly the search process to the most promising part of the solution space.

Exponential Growth of the Number of Extremes. For the mentioned already *Griewank's benchmark* function the number of local extremes grows exponentially with the dimension of the space. For larger size of the space, this fact practically eliminates methods which completely examine even a small fraction of local extremes, and fully disqualifies exhaustive search methods.

Uneven Distribution of Extremes. This phenomenon is well illustrated by *Langermann's benchmark* function which exhibits numerous local minima unevenly distributed, depending on some parameters unknown apriori. It means that strategic search directions in the solution space have no regular character and have to be set in an adaptive way.

Deception Extremes. This phenomenon is well illustrated by *Shakel's benchmark* function (called also *fox holes*) which exhibits quite deep local minima (holes) unevenly distributed on the almost flat surface. The behavior of the function between holes (significant part of the surface) provides no information about minima expected in the vicinity. Moreover, iterative search methods (walking step-by-step) are often unable to go out from so deep minima, which results premature convergence.

Flat Valey of Extremes. This phenomenon is well illustrated by *Rosenbrock's benchmark* function (also known as *banana* function) which owns global minimum located inside a long, narrow, parabolidal flat valley. Finding of the valley is trivial, however convergence of procedure going step-by-step to the global optimum is difficult and very slow.

Curse of Dimensionality. Benchmarks mentioned in three previous heads refer to the space dimension a few or a dozen or so $(1 \ldots 10)$. Nobody have analyzed the behavior of the proper methods for grater space dimension of size of hundreds or thousands (real-case size). In particular, in the paper [15] there was mentioned about *very small* practical instance from scheduling theory with dimensionality 90 and approximately $4 \cdot 10^{48}$ feasible solutions. We are able to check 10^9 solution in practically acceptable time, which constitutes infinitisimaly small fraction 10^{-39} of the whole space. Nobody can enumerate significantly more.

NP Hardness. Most of discrete optimization problems derived from practice are NP-hard, which immediately implies exponential-time computational complexity of solution algorithm. Since the power of processors increases linearly in recent years, while the cost of calculations as well as the number of local extremes increases exponentially with the size of problem, there is no hope to solve real instances in the time acceptable in practice.

Calculation Cost. NP-hardness implies unacceptable large calculation cost measured by the processor running time. Moreover, discrete problems are considered as superfluously rigid, in that sense that small perturbation of data destroys optimality of expensively found solution, which force the user to make expensive calculation once again. That's why seeking optimal solution is not popular in the society of practitioners.

4 Space Landscape

Intuitively, the behavior of the solution algorithm have to be adjusted to the rough landscape of the solution space in order to exploit fully acquired information about its structure. Notion *landscape* is actually an unprecise term since refers to human's intuition of perceving and interpreting 3D view. In fact, solution space is multi-dimensional, thus intuitions such as search directions, trajectory, convexity, have not so intuitive interpretations. Detection of several recognized properties of the landscape (mentioned in the previous section) allow us to design efficient algorithms. Notice, there are at least two views on employing provided knowledge: (1) static, predefined; (2) dynamic, on-line, adaptive. In the approach (1) one can distinguish generally three phases: (A) phase of analyzing (e.g. by sampling) space structure; (B) calculating tuning parameters for solution algorithm; (C) searching solution with current configuration of the algorithm. The approach (2) continously collects information about solution space obtained during the search, which can be used to control searching process in an adaptive way in the on-line mode. Between these two extreme views, there exist a lot of *intermediate* constructions. Particular solution methods realizes these postulates in different ways.

Space Sampling. Sampling can be performed by using *random* overview, or generaing *local search* or *goal oriented search* trajectories walking through the space by neighboring solutions (distant by one unit). It can be performed with respect to a set of any solutions or focusing on local extremes only. It is used to examine or detect space and landscape properties. Random sampling has at least two goals: (1) identifying regions containing feasible solutions, (2) identifying the promising search regions in terms of $K(x)$.

Distributions. Distribution of solutions as well as local minima in the discrete space is usually uneven. One can verify this fact by making space sampling. From such sampling we can find that the distance of any solution to the optimal one is usually distributed normally with the mean about 50% of the space diameter. The distribution of goal function value is also close to normal with the mean depending on the problem considered. Interestingly, the probability of finding solution very close to optimal by random sampling of the space $\mathcal{X}$ is practically infinitisimal, although the number of such solution is so large that there is no way to enumerate them, even partially. This is a serious drawback of random search methods.

Big Valley. Problem is suspected to own big valley phenomenon if there is exist positive correlation (in statistical sense) between goal function value and the distance to optimal solution (the best found solution); in the big valley appears the densification of local extremes of the goal function value. The size of the valley is usually relatively small with respect to the size of the whole solution space.

Ruggedness. This is an independent measure characterizing diversity of goal function values of related (usually neighboring) solutions. Greater ruggedness means sharper and unpredicted changes of $K(x)$ for neighboring solutions. Smaller ruggedness means flat or slow-changeable landscape. There has been proposed objective measures of ruggedness based on autocorrelation coefficient and or autocorrelation function, see [15] for detail.

Other Measures. Among other measures characterizing landscape there are mentioned: correlation between random trajectories, landscape statistically isotropic, fractal landscape, correlation between genes (epitasis), correlation of the distance of fitness.

5 Solution Approaches

The evolution of solution approaches for discrete problems has long and rich history, see Fig. 1. Although milestones of this history presented below deal fundamentally with single-criteria case, they have an influence on multiple-criteria solution methods as well. Beginning from the commonly used *heuristics* based chiefly on various *priority rules* in the fifties and sixties, through the *theory of NP-completeness* (the seventies) which classified problems and algorithms into *polynomial-time* and *exponential-time*. Significant development of *exact algorithms* in the seventies and eighties moved slightly the border of instance sizes which can be solved by these methods but finally set the limit of its applicability. Pessimistic experience with exact methods stimulates, among others, the development of *approximation algorithms* (the eighties and nineties) and *approximality theory*. Besides the theoretical results, a lot of *approximation scheme* (AS), *polynomial-time AS* (PTAS) and *fully polynomial time AS* (FPTAS) were proposed. For the class of on-line algorithms the similar role plays so called *competitive analysis*. However these quite complex theoretical constructions do not

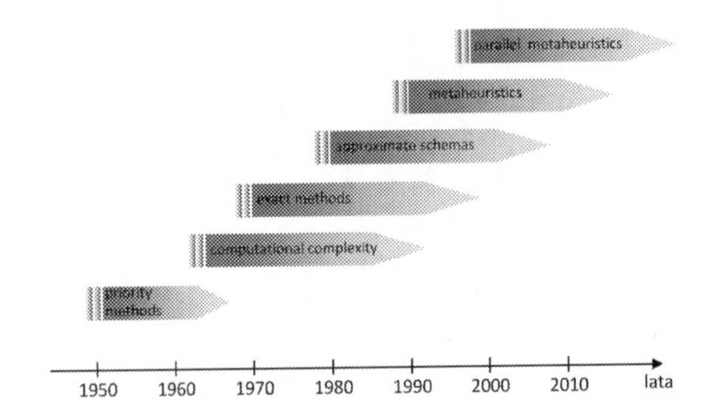

Fig. 1. Development of the solution approaches

gain acceptance among practitioners. From the nineties there was observed the snappy development of *metaheuristics* with good accuracy confirmed in computer benchmarks. Theoretical foundations of metaheuristics appeared a few years later. From 2000, in the natural way, there began the era of meta^{2}heuristics and parallel metaheuristics, being the new class of algorithms.

Exact Methods. We call the method *exact* if it always finds x^* satisfying (1). Depending on the class of computational complexity, one can distinguish the following types of exact methods: (a) dedicated polynomial-time algorithms, (b) algorithms based on the Branch-and-Bound (B&B) scheme, (c) algorithms based on the Dynamic Programming (DP) scheme, (d) algorithms based on Integer Linear Programing (ILP), (e) algorithms based on Binary Linear Programing (BLP), (f) subgradient methods. Methods (a) are considered as computationally cheap specialized methods for problems from P-class or NP-hard numeral. Methods (b) – (f) are computationally expensive, dedicated for strongly NP-hard problems. Up to the end of the eighties one considered them as "sole right" approaches for strongly NP-hard problems, after that time there was appeared barrier of dimension. Although significant development was done in its progress, practitioners still consider them as unattractive, or limit their applications to a narrow scope. Methods are time- and memory- consuming, whereas size of instances which can be solved in a reasonable time is still too small for practice. Moreover, implementation of more complex algorithms of this type needs skillful programmers. The serious problem is also validity of the instance data, which frequently have been perturbed just after the expensive finding of optimal solution and so called superfluous rigidness of the problem. One can say that the cost of finding optimal solution is still to high with the profits obtained from its implementation. Nevertheless, there still exist several problems where application of exact methods are justified and recommended.

Approximate Methods. *Approximate algorithm* A provides solution x^A, so that

$$K(x^A) = \min_{x \in \mathcal{X}^A} K(x) \geq K(x^*) \tag{2}$$

where $\mathcal{X}^A \subset \mathcal{X}$ is the subset of solutions checked by A. The overall aim is to find x^A so that $K(x^A)$ is *close* to $K(x^*)$ by examining the smallest as possible $\mathcal{X}^A$. The closeness to $K(x^*)$ (accuracy) can be either guaranteed a priori or evaluated a posteriori. It is clear that accuracy has opposing tendency to running time, i.e. finding better approximate solution needs longer running time (greater $\mathcal{X}^A$), and this dependence owns strongly nonlinear character. Therefore, discrete optimization manifests a variety of models and solution methods, usually dedicated for narrow classes of problems or even separate problems. Reduction of the generality of models allow us to find special features of the problem, application of which improve numerical properties of the algorithm such that running time, speed of convergence. Quite often, a strongly NP-hard problem has in the literature several various algorithms with different numerical characteristics. Knowledge about models and algorithms allow us to fit satisfactory algorithm for each newly stated problem. Bear in mind, in the considered research area the goal *is not* to formulate whatsoever model and method, but to provide *simply* model and solution method *reasonable* from the computer implementation point of view.

Approximation Error. The set of data specifies the *instance* Z of the problem. Denote by $\mathcal{X}(Z)$ the set of feasible solutions for the problem, and by $K(x; Z)$ value of criteria K for solution x in the instance Z. Solution $x^* \in \mathcal{X}(Z)$ such that $K(x^*; Z) = \min\{K(x; Z) : x \in \mathcal{X}(Z)\}$ is the optimal solution for the instance Z. Let $x^A \in \mathcal{X}(Z)$ denote the approximate solution generated by algorithm A for the instance Z. The approximation error $F^A(Z)$ of algorithm A observed on instance Z is a measure defined on the base of relation between $K(x^A; Z)$ and $K(x^*; Z)$, for example $F^A(Z) = S^A(Z) \stackrel{\text{def}}{=} K(x^A; Z)/K(x^*; Z)$, see [] short survey for other definitions of $F^A(Z)$. Behaviour of $F^A(Z)$ over Z can be examined either experimentally or analytically, apriori or a posteriori.

Experimental Analysis. It evaluates *a posteriori* behavior of the algorithm (chosen error, running time, etc.) on the base of results obtained for limited *representative sample* of instances Z. This is the most popular method despite its main drawback, namely it depends on the chosen sample of instances (is subjective). However, only this analysis is able, to justify, in the context of "no free lunch" theorem, observed superiority of the chosen algorithm over other subclasses of instances Z.

Worst Case Analysis. It evaluates *a priori* behavior of the chosen error F on *entire* infinite population of instances Z. Usually, there is applied for mentioned already error $F^A(Z) = S^A(Z) = K(x^A; Z)/K(x^*; Z)$, for which there are also defined the *worst-case ratio* $\eta^A = \min\{y : S^A(Z) \leq y, \ \forall Z\}$ and *asymptotic worst-case ratio* $\eta^A_\infty = \min\{y : S^A(Z) \leq y, \ \forall Z \in \{W : K(x^*; W) \geq L\}\}$, where L is a number.

Probabilistic Analysis. It evaluates *a priori* behavior of the chosen error F on *entire* infinite population of instances Z, considering each instance Z as a realization of n independent random variables with known distributions (usually uniform) of probability; this fact will be denoted by writing Z_n instead of Z. Then, values $K(x^*; Z_n)$, $K(x^A; Z_n)$ and $F^A(Z_n)$ are clearly random variables. The analysis provides basic information about behavior of random variable $F^A(Z_n)$, namely its distribution, moments, etc. and *type* and *speed and type of convergence* with n tending to infinicity. For example, there are considered the following types of convergence: (a) almost sure $P(\lim_{n\to\infty} F^A(Z_n) = m) = 1$, (b) in probability $\lim_{n\to\infty} P(|F^A(Z_n) - m| > \epsilon) = 0$, for any $\epsilon > 0$, (c) in the mean $\lim_{n\to\infty} |E(F^A(Z_n)) - m| = 0$.

Approximation Schemes. Approximation scheme (AS) is the family of algorithms A, such that A provides for the given $\epsilon > 0$ solution x^A satisfying $K(x^A; Z)/K(x^*; Z) \leq 1 + \epsilon, \quad \forall Z$. AS is the *polynomial-time approximation scheme* (PTAS), if for any fixed ϵ it owns polynomial computational complexity. If additionally this complexity is a polynomial of $1/\epsilon$, then scheme is *fully polynomial-time approximation scheme* (FPTAS). In practice, ASes turned out to be rather complex algorithmic constructions and appear inactractive for practical applications.

6 The Newest Trends

In recent years, simultaneously with the development of mathematically perfect theories, there has been observed rapid development of *metaheuristics*, i.e. approximate methods without excessive theory but with good or even excellent numerical properties confirmed in numerous computer tests. Surprisingly, these methods are more interesting for users, since in practice provide quickly solutions with better quality, than mathematically perfect approximation schemes. These methods are classified as either *constructive* (see first two entries in Table 1) or *improvement* (see the remain entries in Table 1). The former are fast, easily implementable, but generate solutions of poor quality. The later are slower, need starting solution improved next iteratively, but provide solutions with good or excellent quality. They also allow to form in a flexible way the compromise between the solution quality and the algorithm's running time. Theoretical guarantee of quality were found, up till now, for numerous constructive methods but only for few improvement methods. For some improvement methods there have been proved convergence to the optimal solution, the sufficient conditions do not hold in practice, thus these results have rather theoretical then practical significance. Finally, the practical usefulness of approaches and/or algorithms follows from various theoretical as well as experimental analysis.

7 Multicriteria Approaches

Practitioners usually evaluate solutions taking into account various points of view, thus using a number of different criteria. Thus, in this section we consider the following optimization problem: find $x^* \in \mathcal{X}$ such that

$$K^* \stackrel{\text{def}}{=} K(x^*) = \min_{x \in \mathcal{X}} K(x), \tag{3}$$

Table 1. List of metaheuristic approaches

Aproaches
Constructive algorithms (CA)
Priority rules (PR)
Local search (LS)
Descending search, hill climbing (DS)
Random search (RS), Monte Carlo methods (MC)
Simulated annealing (SA)
Simulated jumping (SJ)
Tabu search (TS)
Adaptive memory search (AMS)
Path search, star path search (PS, SPS)
Goal oriented tracing paths (GOTP)
Curtailed (truncated) branch-and-bound (CB&B)
Randomised methods (RM)
Greedy random adaptive search procedure (GRASP)
Variable neighborhood search (VNS)
Beam search (BS), filtered beam search (FBS)
Guided local search (GLS)
Genetic, evolutionary search (GS)
Memetic search (MS)
Differential evolution (DE)
Cultural methods (CM)
Artificial immune system (AIS)
Path relinking (PR)
Biochemical random search (BRS)
Ant colony optimization (ACO)
Scatter search (SS)
Constraint satisfaction (CS)
Geometric approach (GES)
Particle swarm optimization (PSO)
Bee search (BS)
Bat search (BA)
Harmony search (HS)
Electromagnetic search (ES)
Intelligent wather drops (IWD)
Neural nets (NN)

where

$$K(x) = [K_1(x), \ldots, K_s(x)]^T \tag{4}$$

and x, x^*, $\mathcal{X}$ and $K(x)$ are solution, optimal solution, set of feasible solutions and vector goal function, respectively. The min operator in (3) does not specify how to interprete minimization over the set of vectors since formally we need to define the method of comparison between vectors which depends on user preferences expressed directly or undirectly. The primary goal of multiobjective optimization is to model preferences of the decision maker (expresses as the importance of each particular criteria or ordered rank of criteria).

Known multicriteria solution approaches are classified depending on the philosophy of expressing user's preferences: (1) preferences are defined by the user *a priori* as relative importance of the component criteria, (2) user expresses preferences *a posteriori* by selecting one solution from the set of uncomparable (equivalent) solutions, (3) no preferences are provided by the user, (4) preferences are set in certain iterative way (progressively) by learning equally the system and user how to find the satisfactory solution. Special attension is set to (5) genetic population-based approaches in the context of finding Pareto frontier.

There exists also another classification, which distinguishes basically two classes of methods: (a) optimization by scalarization, (b) pure vector optimization methods. The former approach is clear since, by using a transformation, leads to single-optimization case, which provides single optimal solution in terms of combined function. For such problems, results presented in Sections 2 – 6 remains valid. In the latter approach there is no single global solution, but there is a set of solutions that satisfy so called Pareto optimality (in strong or weak sense).

Skipping the formal definition of Pareto optimality we only mention about valid topics, methods and notions associated with this subject, namely: (A) necessary and sufficient conditions for solution to be Pareto optimal, (B) method of checking whether given solution is Pareto optimal, (C) notion of efficient and dominated/undominated solutions, (D) notion of compromise solution, (E) notion of utopia point, (F) non-dimensional objective transformations.

Follow the approach (1) one can find in the literature a lot of scalarizing function, see survey in [13], which lead to particular methods known as: (1.1) weighted global criterion method, (1.2) weighted sum method, (1.3) lexicographic method, (1.4) weighted min-max method, (1.5) exponential weighted criterion, (1.6) weighted product method, (1.7) goal programming methods, (1.8) bounded objective function method, (1.9) physical programming. The scalarizing techniques are wide, begining for example from a simple (1.2) $K(x) = \sum_{i=1}^{s} w_i K_i(x)$ with arbitrary or user-defined weights w_i, up to quite sophisticated (1.4) $K(x) = \max_i\{w_i[K_i(x) - K_i^o]\} + \rho \sum_{j=1}^{s} w_j[K_j(x) - K_j^o]$, where K^o denotes utopia point. For more details we refer the reader to the paper [13].

In the approach (2) the algorithm provides a representation of the Pareto optimal set (or subset) leaving the final decision for the decision maker. The basic aim of this approach is to produce a set of optimal Pareto points which are able to represent acurately the complete Pareto set. Among particular methods one can find: (2.1) physical programming, (2.2) normal boundary intersection, (2.3) normal constraint. Notice, these are rather expensive technologies providing the Pareto frontier or its approximation.

If the decision-maker cannot define her/his preferences, methods from group (3) is recommended. In practice, approaches from group (1) are applied with the exclusion of user-defined parameters. In this context, respecive algorithms can be used to group (3) as well. Careful study of this area allow us to distinguish the following methods: (3.1) global criterion (i.e. scalarization with arbitrary parameter values), (3.2) Nash arbitration and objective product, (3.3.) Rao's method.

Decision maker preferences set in an iterative way are usually modelled as special class of games. They commonly are used in Decision Support Systems based on Muliple Criteria Decision Making (MCDM), see Table 2 for known aprooaches in this area.

Special attention has been paid to genetic algorithms (GA) due to their particular usefulness in solving multicriteria problems since they naturally operate on the set of dispersed solutions (population). This technology is clear in the context of single criteria case and for multicriteria case with a scalarization of the goal function, see previous sections. The most interesting is the pure vector optimization, for which GA is able to provide quite efficiently an aproximation of Pareto frontier, see the review in [11]. Up to now, several various original approaches were developed in this area. Skipping consciously the overview of these approaches we mention only about some particular methods: (5.1) Weighted-sum-Approach (using randomly generated weigths and Elitism), (5.2) Vector Evaluated Genetic Algorithm (VEGA), (5.3.) Multi-Objective Genetic Agorithm (MOGA), (5.4) Nitched Pareto Genetic Algorithm (NPGA), (5.5) Strength Pareto Evolutionary Algorithm (SPA), (5.6) Non-dominated Sorting Genetic Algorithm (NSGA), (5.7) Vector-optimized evolution strategy (VOES), (5.8) Weight-based genetic algorithm (WBGA), (5.9) Predator-prey evolution strategy (PPES), (5.10) Elitist multi-objective evolutionary algorithm (EMOEA), (5.11) Elitist non-dominated sorting genetic algorithm (ENSGA), (5.12) Distance-Based Pareto genetic algorithm (DBPGA), (5.13) Thermodynamical genetic algorithm (TGA), (5.14) Pareto-archived evolution strategy (PAES).

8 Parallel Methods

In recent years the increase of computational power of computers evolves towards parallel architectures. Since the increase of the number of processors or cores in single computer is still too slow comparing it with the increase of the number of solutions in the space, there is no hope to vanquish barrier of NP-hardness in the area of exact methods. Even cloud computing with the use of computer clusters does not offer good alternative, chiefly because of too high calculation cost. On the other hand, computer parallelism can improve significantly metaheuristics in terms of running time and quality. Thus parallel metaheuristics become the most desired class of algorithms, since they link excellent quality with a short running time. Sophisticated implementations of parallel algorithms require skilfull application of a few fundamental elements linked with parallel programming theory, calculation models, and practical tools, namely: (1) theoretical models of parallel calculation (SISD, SIMD, MISD, MIMD), (2) theoretical models of memory access (EREW, CREW, CRCW), (3) practical parallel calculation environments (hardware, software, GPGPU), (4) shared memory programming (Pthreads in C,

Table 2. List of MCDM approaches

Aproaches
Aggregated Indices Randomization Method (AIRM)
Analytic hierarchy process (AHP)
Analytic network process (ANP)
Data envelopment analysis
Decision EXpert (DEX)
Dominance-based rough set approach (DRSA)
ELECTRE (Outranking)
The evidential reasoning approach (ER)
Goal programming
Grey relational analysis (GRA)
Inner product of vectors (IPV)
Measuring Attractiveness by a Cathegorial Based Evaluation Technique (MACBETH)
Disaggregation – Aggregation Approaches (UTA*, UTAII, UTADIS)
Multi-Attribute Global Inference of Quality (MAGIQ)
Multi-attribute utility theory (MAUT)
Multi-attribute value theory (MAVT)
New Approach to Appraisal (NATA)
Nonstructural Fuzzy Decision Support System (NSFDSS)
Potentially all pairwise rankings of all possible alternatives (PAPRIKA)
PROMETHEE (Outranking)
Superiority and inferiority ranking method (SIR)
Technique for the Order of Prioritisation by Similarity to Ideal Solution (TOPSIS)
Value analysis (VA)
Value engineering (VE)
VIKOR method
Fuzzy VIKOR method
Weighted product model (WPM)
Weighted sum model (WSM)

Java threads, Open MP in FORTRAN, C, C++), (5) distributed memory programming, message-passing, object-based, (6) Internet computing (PVM, MPI, Sockets, Java RMI, CORBA, Globus, Condor), (6) measures of quality of parallel algorithms (runtime, speedup, efficiency, cost), (7) single/multiple searching threads, (8) granularity evaluation, (9) independent/cooperative search threads, (10) distributed (reliable) calculations in the net.

It is noticeable, that metahuristics can be implemented in parallel calculation environments in different manner, providing particular algorithms with various numerical properties. Let us consider, for example, SA approach. We can adopt this method as follows: (a) single thread, conventional SA, parallel calculation of the goal function value, fine grain, conventional theory of convergence,

(b) single thread, pSA, parallel moves, subset of random trial solutions selected in the neighborhood, parallel evaluation of trial solutions, parallel theory of convergence, (c) exploration of equilibrium state at fixed temperature in parallel, (d) multiple independent threads, coarse grain, (e) multiple cooperative threads, coarse grain. Similarly, for GS we have: (a) single thread, conventional GA, parallel calculation of the goal function value, small grain, theory of convergence, (b) single thread, parallel evaluation of population, (c) multiple independent threads, coarse grain, (d) multiple cooperative threads, (e) distributed subpopulations, migration, diffusion, island models. These means that from several sequential methods we can create many parallel methods, so the final number of possible solution technologies is quite large.

9 Optimization Strategies

Neither (1) nor (3) define precisely the method of calculating $K(x)$ for the given x. Notice, because of the hardness of the most practical optimization tasks, one can expect that the optimization procedure uses an approximate searching strategy A, which for single-criteria case seeks solution x^A so $K(x^A) = \min_{x \in \mathcal{X}^A} K(x)$ by immediate checking values of the goal function $K(x)$ only for some subset $x \in \mathcal{X}^A \subset \mathcal{X}$, see formula (2). In the multiple criteria case we select among solutions from $\mathcal{X}^A$ the set of undominated solutions, providing in this way certain approximation of Pareto front. The cost of such calculations depend on the cardinality of $\mathcal{X}^A$ and the computational complexity of performing the basic step "for the given x find $K(x)$". In case of too high cost of calculations, one can either replace $K(x)$ by a cheapest its approximation $K'(x)$ or by limiting cardinality of $\mathcal{X}^A$. It is evident that these two elements (namely $\mathcal{X}^A$ and cost of $K(x)$) correlate and infuence strictly not only on the final result provided by the solution algorithm but also on technology of designing such algorithm. Moreover, the computational complexity depends also on the character of the data provided to the optimization task. After an analysis we propose the following taxonomy:

- x is deterministic, function $K(x)$ is given by a formula (clear, the most frequent case),
- x is deterministic, function $K(x)$ is given by a deterministic polynomial-time algorithm (e.g. longest path in the graph defined by x),
- x is deterministic, function $K(x)$ is given by a deterministic exponential-time algorithm (e.g. TSP for given set of cities x),
- x is deterministic, function $K(x)$ is given by a deterministic algorithm provided in form of pseudocode or program code,
- x is random variable, function $K(x)$ represents certain measure on x (e.g. moments, probability),
- x is random variable, function $K(x)$ is given by an algorithm (e.g. RR in simple queuening systems),
- x is fuzzy variable, function $K(x)$ represents certain defuzzified measure on x,

– x is any variable, function $K(x)$ is given as the result of running program code (especially the result of a simulation),
– x is any variable, function $K(x)$ is given as the result of sensor measurement.

The third case clearly show that in some situations we have to replace the exact calculation of $K(x)$ for the given x by some approximation denoted hereinafer $K'(x)$. Two last enumerated cases can be treated as experimental measurement of the black box, see Figure 2, with the input x and the output $K(x)$. Notice that

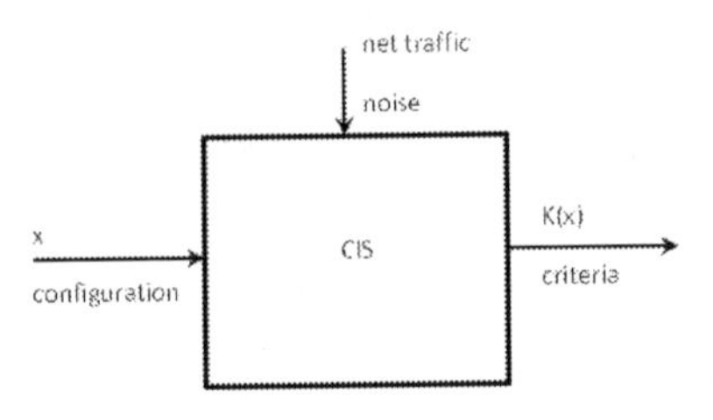

Fig. 2. CIS as an object of control

the same input values may provide different output values because of the noise. Therefore, we are incline to the view that based on the sequence of measurements for the same x we need to define certain estimation of the result $K(x)$.

10 Proposed Approach

Refeering to the taxonomy provided in Section 9 one can say that the form of optimization task for the case of CIS collapse depends on the style of CIS description. Taking into account fundamental features of the CIS architecture and activity, the simulation seems to be the most adequate method of $K(x)$ calculation, see Fig. 3. Effects of such approach are manifold. First, solution x corresponds to configuration of services in CIS, i.e. their distribution among nodes. The service i located at the node j using contrary policy of queue or resources is treated as different solution. Requirements coming to CIS from workstations are treated as the noise from statistical point of view. Single simulation provides a measurement of some parameter(s) representing criteria $K(x)$ treated as observed realization of the random variable. Statistically important sample length is necessary to estimate correctly output of our black box being the object of control. Box called "x control" is in fact one among mentioned earlier solution methods, e.g. GA, SA, etc.

11 Attacks and Malfunctions

Malfunctions in CIS depends on the connections (links) in the net as well as on the availability of services located in nodes of the network. In order to ensure

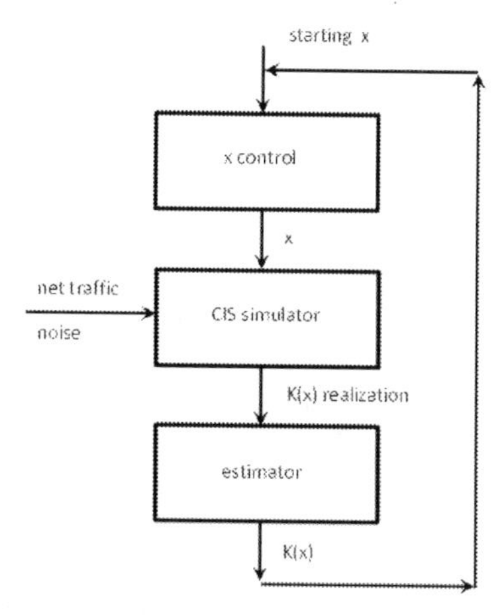

Fig. 3. Proposed optimization algorithm

level of protection some services can be redundant and available in a few nodes by using various paths. Each failure in CIS entails several scenarios (choreography) of changes, which aimed to restoration functionalities/availability of services. These choreographies are either given a priori (fixed) or generated automatically by some generator. In [8] there has been provided some classifications of failures, which can occur in the network, namely by: (a) scope of protection, (b) backup path setup method, (c) type of resource reservation, (d) domain of recovery process. Note that due to statistical character, failures usually appear in single point (node) at once. It is commonly assumed that probability of occuring failure in several nodes simultaneously is close to zero. Separate case to discuss and consider is so called progressive failures. Refering to attacks, we distinguish attacks carried on on single node, however attacks carried on on multiple nodes are possible as well.

12 Criteria

As a result of malfunction some services become unavailable. The basic aim of reconfiguration is to restore full functionality of CIS as quick as possible. To this order several choreografies are possible and one of them is chosen for application. All of them are evaluated from several points of view. Thus, among considered criteria one can use: (1) maximal restoration time, (2) average restoration time, (3) aggregate restoration time, (4) fraction of unserviced clients, (5) rejection ratio, (6) resource utilization, (7) cost of restoration.

13 Topologies

Simulation of choreography can be performed on CIS taken from reality assuming different occuring malfunctions. However, this provides only single or few

Table 3. Parameters characterizing topologies

Parameters
Betweenness centrality
Clustering coefficient
Global efficiency of a graph
Local efficiency
Small world property
Scale-free property
Rich-club connectivity of a graph
Status of a vertex, graph
Median of a graph
Centroid value of a vertex, graph
Centroid of a graph
Normalized Laplacian spectrum of a graph
Maximal fault tolerance of a graph
Maximum flow

instance(s) of CIS. Since the proposed approach provides only an approximate solution, then in order to evaluate overall its features more testbeds need to be used, see Section 5 for methodology of approximate algorithms analysis. To this aim, some generators of the net topologies are especially welcome, particularly for scale-free networks like Internet. To distinguish various topologies a lot of parameters has been used in the literature, see Table 3 based on the survey in [8]. These parameters follow from practice and constitutes the base for net generators for the simulator. One can find a few deterministic topologies: (a) circulant graphs lattices, (b) chordal rings, (c) Cayley graphs, (d) hypercubes, and a lot of non-deterministic topologies, called as following models: (e) random graph (of Erdö and Rényi, ER), (f) Watts-Strogatz (WS), (g) Waxman (WX), (h) CRUG, (i) C-CRUG-PA, (j) C-CRUG-MAX-DPA, (k) NPART. The most suitable are generators of Internet-like topology, namely models: (l) Barabasi Albert (BA), (m) Extended Barabasi Albert (EBA), (n) Tiers (TI), (o) Transit-Stub (TS), (p) Power Law Random Graph (PLRG), (q) BRITE, (r) Interactive Growth (IG), (s) Positive Feedback Preference (PFP), (t) Inet-3.0 (INET).

Not discessing here in details each particular model, we only mention that the result of simulation depends on chosen model of the net. For our aim the free-scale network generator similar to Internet is recommended. Finding optimal strategy of reconfiguration needs excesive computational experiments.

14 Conclusions and Comments

The given survey of methodologies leading to the proposed CIS optimization task does not provide all details necessary to make an algorithm. It rather outlines crucial aspectcs important for the design and context of use of solution methods for the hard discrete optimization problems in the environment having rich variety of possible approaches. The present tendency prefer metaheuristics (sequencing as well as parallel, also in multicriteria case) since they links high or good quality of generated solutions with relatively small or moderate calculation cost. Morover they are resistant to local extremes. Real usefulness and applicability of each particular method depends on space landscape, rutherness, big valley, distribution of solutions in the space and the probler balance between instesification and diversification of the search. Currently, for the single criteria case, the promissing approaches are SA, SJ, GS, MS – for problems without any particular properties (SA and SJ for problems having high cost of evaluating single solution) and TS, AMS – for problems having special properties allowing on acceleration of the searching process. For the multicriteria case recommended are population based methods, namely GS, ACO, or methods that operate on the sets of numerous solutions like TS. Recent study suggest that eficcient finding of Pareto from can be done by united force of a few different algorithms, e.g. GA+ACO+TS. If cost of calculations becomes high, for example for instances of greater size, there is recommended to consider parallel methods, possible to implement already on a PC with multicore processor or CUDA platform.

Coming back to the task of calculating $K(x)$ for the given x via simulation, we suggest to give up seeking Pareto front or its approximation due to very high cost of calculations. Taking into account this unusual technology of $K(x)$ finding, a scalarization is the most appropriate approach and SA is the most recommended metaheuristics.

Acknowledgments. Paper is supported by funds of National Centre of Science, agreement 4759/B/T02/2011/40, grant N N516 475940.

References

1. Aarts, E.H.L., van Laarhoven, P.J.M.: Simulated annealing: a pedestrain review of the theory and some applications. In: Deviijver, P.A., Kittler, J. (eds.) Pattern Recognition and Applications. Springer, Berlin (1987)
2. Alba, E.: Parallel metaheuristics: a new class of algorithms. John Wiley & Sons (2005)
3. Bartak, R.: On-line guide to Constraint programming (2010), `http://ktiml.mff.cuni.cz/bartak/constraints/`
4. Bożejko, W.: A new class of parallel scheduling algorithms. Oficyna Wydawnicza PWr, Wrocław (2010)
5. Corne, D., Dorigo, M., Glover, F. (eds.): New ideas in optimization. McGraw Hill, Cambridge (1999)
6. Dorigo, M., Stützle, T.: Ant Colony Optimization. Bradford Books (2004)
7. Geem, Z.W. (ed.): Recent Advances in Harmony Search Algorithm. SCI, vol. 270. Springer, Heidelberg (2010)
8. Gierszewski, T.: Methods for minimizing attack's impact on IP networks, PhD Dissertation, Gdansk University of Technology, Gdansk, Poland (2011)
9. Glover, F., Laguna, M.: Tabu search. Kluwer Academic Publishers, Boston (1997)

10. Goldberg, D.E.: Genetic algorithms in search, optimization and machine learning. Addison-Wesley (1989)
11. Ghosh, A., Dehuri, S.: Evolutionary Algorithms for Multi-Criterion Optimization: A Servey. International Journal of Computing & Information Sciences 2(1), 38–57 (2004)
12. Nedjah, N., Coelho, L.S., de Mourelle, L.M. (eds.): Multi-Objective Swarm Intelligent Systems. SCI, vol. 261. Springer, Heidelberg (2009)
13. Marler, R.T., Arora, J.S.: Survey of multi-objective optimization methods for engineering. Struct. Multidisc. Optim. 26, 369–395 (2004)
14. Molga, M., Smutnicki, C.: Test functions for optimization needs, Technical Report, Institute of Computer Engineering Control and Robotics, Wroclaw University of Technology, Wroclaw, Poland (2005)
15. Smutnicki, C.: Optimization technologies for hard problems. In: Fodor, J., Klempous, R., Suárez Araujo, C.P. (eds.) Recent Advances in Intelligent Engineering Systems. Studies in Computational Intelligence, vol. 378, pp. 79–104. Springer, Heidelberg (2012)
16. Tartanoglu, F., Issarny, V., Romanovsky, A., Levy, N.: Dependability in the Web Service Architecture. In: de Lemos, R., Gacek, C., Romanovsky, A. (eds.) Architecting Dependable Systems. LNCS, vol. 2677, pp. 90–109. Springer, Heidelberg (2003)
17. Wierzchon, S.T.: Artificial immune systems. Theory and application, EXIT, Warsaw (2001) (in Polish)

Metascheduling Strategies in Distributed Computing with Non-dedicated Resources

Victor Toporkov[1], Alexey Tselishchev[2], Dmitry Yemelyanov[1], and Petr Potekhin[1]

[1] National Research University "MPEI",
ul. Krasnokazarmennaya 14, Moscow, 111250 Russia
{ToporkovVV,YemelyanovDM,PotekhinPA}@mpei.ru
[2] CERN (European Organization for Nuclear Research),
CERN CH-1211 Genève 23 Switzerland
Alexey.Tselishchev@cern.ch

Abstract. In this chapter, we address problems of efficient computing in distributed systems with non-dedicated resources including utility Grid. There are global job flows from external users along with resource owner's local tasks upon resource non-dedication condition. Competition for resource reservation between independent users, local and global job flows substantially complicates scheduling and the requirement to provide the necessary quality of service. A metascheduling concept, justified in this work, assumes a complex combination of job flow dispatching and application-level scheduling methods for parallel jobs, as well as resource sharing and consumption policies established in virtual organizations and based on economic principles.

Keywords: Distributed computing, economic scheduling, resource management, co-allocation, slot, job, task, batch.

1 Introduction

Execution of large parallel jobs in distributed computational environments requires allocation of significant resources amount partially shared with their owners [1-4]. Today well-known algorithms, their combinations and heuristics used by schedulers are usually unable to provide optimal or suboptimal solutions in terms of heterogeneous distributed environments and dynamically changing sets of available computational nodes and their utilization. Resource management and job scheduling economic models proved to be efficient in such conditions [1-3].

Two established trends may be outlined among diverse approaches to distributed computing. The first one is based on the available resources utilization and application-level scheduling. As a rule, this approach does not imply any global resource sharing or allocation policy. Application agents, i.e. resource brokers [5-11], are usually considered as mediators between the users and the resource owners. There are a lot of projects belonging to this trend, namely AppLeS [6], APST [7], Legion [8], DRM [9], Condor-G [10], Nimrod/G [11] and others.

W. Zamojski and J. Sugier (eds.), *Dependability Problems of Complex Information Systems*,
Advances in Intelligent Systems and Computing 307, DOI: 10.1007/978-3-319-08964-5_8

Another trend is related to the formation of user's virtual organizations (VO) and job flow scheduling [12-14]. In this case, an external scheduler, e.g. a Grid dispatcher, a metascheduler or a Meta-Broker [15], is an intermediate chain between the users and local resource management and job batch processing systems.

Scheduling and resource management systems belonging to the first approach are well-scalable and application-oriented. However, simultaneous application-level scheduling with diverse optimization criteria set by independent users, especially upon possible competition between applications, may deteriorate such integral QoS characteristics of a distributed environment as total job batch execution time or overall resource utilization. VOs, from one hand, naturally restrict the scalability of resource management systems. On the other hand, uniform rules of resource sharing and consumption, in particular based on economic models [1-4, 16-18], makes it possible to improve the job-flow level scheduling and resource distribution efficiency.

The "convergence" idea of application-level and job-flow scheduling approaches was declared in relatively early works [14, 19-21]. Nevertheless, in some well-known models of distributed computing with non-dedicated resources, only the first fit set of resources is chosen depending on the environment state [22-24], while job scheduling optimization mechanisms are usually not supported. The aspects related to the specifics of environments with non-dedicated resources, particularly dynamic resource loading, the competition between independent users, users' global and owners' local job flows, are not presented in other models [14, 16, 17].

A metascheduling concept in VOs proposed in this work fundamentally differs from known solutions by combining methods of independent job flow management and application-level scheduling [19-21]. We propose a model of independent job flows management based on economic principles The job scheduling is performed cyclically for alternative sets of preliminary selected resources (alternatives) [25]. In contrast to well-known models, the proposed approach assumes job flows and batches formations according to job features, characteristics, resource requirements, users' preferences, and further job batch cyclic scheduling based on dynamically updated VO policies, strategies and restrictions. Job batch schedules are optimized by a criteria vector according to the resource sharing and consumption policy established in the VO.

The rest of this chapter is organized as follows. Section 2 is devoted to analysis of various VO stakeholders preferences and related works in distributed computing. There is a formal problem statement for a cyclic scheduling scheme. Then we discuss restrictions of this scheme. In Section 3, we introduce main requirements for a model of scheduling and fair resource sharing, representing the cyclic scheduling scheme generalization. A combined scheduling approach based on generalized cyclic scheduling scheme and backfilling is proposed in Section 4. Section 5 contains a simulation framework description, variables and parameters for the model of scheduling and fair resource sharing studies. The simulation results are presented in Section 6. Section 7 focuses on the research of the scheduling method combined with backfilling. Finally, Section 8 summarizes the chapter and describes further research topics.

2 Scheduling Problems in VO

2.1 VO Stakeholders and Their Preferences

The scheduling efficiency in VO may be considered from different points of view. On the one hand, one of the most important indicators is available resources utilization level and an average job starting time ("response" time). Computational nodes of distributed environments with non-dedicated resources are generally partially utilized by local high priority tasks. Thus, the available resources of VO are represented as a set of slots, i.e. time spans during which the related node is idle and ready for executing a part of a parallel job. The existence of an available slots set with different start and finish times as well as a different performance rate (depending on the CPU node characteristics), complicates the problem of efficient resource co-allocation and job-flow execution in the distributed environment. The resource fragmentation also reduces the overall distributed environment utilization level. On the other hand, the VO scheduling efficiency may be considered in terms of compliance with certain scheduling policies and VO shareholders preferences. Besides, there are contradictory interests of VO users, resources owners and administrators. The users are usually interested in the earliest start time for their applications with the lowest cost, while resource owners intend to obtain the maximum profit for providing their resources in VO. The administrators define VO policy and they are interested in the distributed environment overall performance optimization as well as in matching preferences of users and resource owners. A fair resource sharing implies that the interests of VO shareholders are met.

Every user job is submitted with a resource request – a list of requirements for the resources needed for a particular application execution. One of the most important parameters is a resource reservation time, during which the allocated nodes are utilized by the user job. For the overall job-flow execution optimization and a resource occupation time prediction existing schedulers rely on the time specified in the job's resource request. However, the reservation time is usually based on user inaccurate runtime estimates [14, 26]. In case, when the application is completed before the term specified in the resource request, the allocated resources remain underutilized. Moreover, if the job runtime estimation substantially differs from the real runtime, the job schedule may become ineffective in terms of optimization criteria defined in VO.

Thus, we outline two main job-flow optimization directions in the distributed computing environment. In the first of them, the optimal or suboptimal scheduling under a given criterion or criteria specified in VO, is performed on the basis of a priori information about local schedules of computational nodes and the resource reservation time for each job execution. The cyclic scheduling scheme (CSS) [27] belongs to this type of systems. Another approach represents scheduling "on the fly" depending on a dynamically updated information about resource utilization. In this case, schedulers are focused on overall resources load maximization and job start time minimizing. Backfilling [28] may be related to this type of scheduling. Existing scheduling approaches are discussed in the next subsection.

2.2 Related Works

There are several resource selection and scheduling algorithms for parallel jobs in distributed environments [17, 22-24, 30-33]. The scheduling problem in Grid is NP-hard due to its combinatorial nature and many heuristic-based solutions have been proposed. In [17], heuristic algorithms for slot selection, based on user-defined utility functions, are introduced. NWIRE system [17] performs a slot window allocation based on the user defined efficiency criterion under the maximum total execution cost constraint. However, the optimization occurs only on the stage of the best found offer selection.

The paper [30] presents architecture and an algorithm for performing Grid resources co-allocation without the need for advance reservations based on synchronous queuing (SQ) of subtasks. The objective of SQ is to minimize the co-allocation skew of all tasks requiring co-allocation. It enables SQ to over subscribe the resources and hence to improve resource utilization. Mean utilization value is a single criterion in this model. However, advance reservation is effective to improve the co-allocation QoS. Moreover job control and resource management may be efficient using strategies. This means a combination of different algorithms and scheduling heuristics [3, 17, 22-24, 27, 32] with consideration for multiple factors and criteria: the policy of resource allocation and administration, dynamical composition and heterogeneity of CPU nodes, etc. By combining the optimization criteria, VO administrators and users can form alternative search strategies for every job in the batch [27]. Users may be interested in their jobs total execution cost minimizing or, for example, in the earliest possible jobs finish time, and in being able to affect the set of alternatives found by specifying the job distribution criteria. VO administrators in turn are interested in finding extreme alternatives characteristics values (e.g., total cost, total execution time) to form more flexible and, possibly, more effective combination of alternatives representing a batch execution schedule.

Advance reservation-based co-allocation algorithms are proposed in [22-24, 31, 32]. First fit resource selection algorithms (backtrack [22, 23] and NorduGrid [24] approaches) assign any job to the first set of slots matching the resource request conditions without any optimization. The co-allocation algorithms described in [31-34] suppose an exhaustive search and some of them are based on a linear integer programming (IP) [3, 32] or mixed-integer programming (MIP) model [33]. In [31] an online algorithm for co-allocating resources that provides support for advance reservations is proposed. The overall complexity of the algorithm for a successful scheduling attempt for the temporal space including a set of Q slots is $O\left(n_r \times Q \times (log\,M)^2\right)$, where M is the number of servers in a computing system, and n_r is the reservation spatial size, i.e., the number of servers required for the given job. The co-allocation algorithm presented in [32] uses the 0-1 IP model with the goal of creating reservation plans satisfying user resource requirements. Users can specify a time frame for each resource: the earliest start time, the latest start time, and the job duration, where the user wants to reserve a time slot. This condition imposes restrictions for slots search only within this time frame. Moreover, the important factor is a complexity and an actual calculation time of the algorithm under consideration [32] especially with the assumption of the repeated use during the scheduling interval. The number of variables in the proposed algorithm becomes R^3 depending on the number of computer sites R. Thus, this approach may be inadequate for an on-line service in practical use. A linear IP-driven algorithm is proposed in [3]. It combines the capabilities of IP and genetic algorithm and allows to obtain the best metaschedule that minimizes the combined cost of all independent users in a coordinated manner. In [33], the authors propose a MIP model which determines the best scheduling for all the jobs in the queue in environments composed of multiple clusters that act collaboratively.

Backfilling [28] is a FCFS (First Come – First Served) method modification. In contrast to FCFS, backfilling requires user's jobs runtime estimates in order to reserve resources in advance. The resources are assigned to the jobs in a priority order, and the jobs are allocated on to the suitable resources if they are not already reserved for higher priority jobs. The advance reservation mechanism in backfilling guarantees to get the resources for higher priority jobs and allows the job queue order violation, which contributes to a higher overall resource utilization. The queue order violation occurs during the backfill stage when low priority jobs are attempted to be allocated to unreserved resources. With backfilling conservative variation a low priority job may be executed out of order, if it will not delay the execution of all higher priority jobs. Aggressive backfilling variation allows jobs to be executed out of the order only in case, when they do not delay the highest priority job execution.

There are some limitations of backfilling for distributed computing. The first one is inefficient resource usage by criteria differed from an average job start time (especially at a relatively low level resources load). The second is a principal inability to affect the resource sharing quality by defining policies and criteria in VO. Nevertheless it is appropriate to consider the use of backfilling to avoid resources fragmentation (see subsection 2.3).

The scheduling techniques proposed in [3, 31-33] are efficient compared with other scheduling techniques under given criteria: the minimum processing cost, the overall makespan, resources utilization etc. However, complexity of the scheduling process is extremely increased by the resources heterogeneity and the co-allocation process, which distributes the tasks of parallel jobs across resource domain boundaries. The degree of complexity may be an obstacle for on-line use in large-scale distributed environments.

In this work, we use algorithms for efficient slot selection based on user and VO administrators defined criteria with the linear complexity on the number of all available time-slots during the scheduling interval denoting how far in the future the system may schedule resources [25, 27, 29]. Besides, in our approach the job start time and the finish time for slot search algorithms may be considered as criteria specified by users in accordance with the job total allocation cost. It makes an opportunity to perform more flexible scheduling solutions.

2.3 Cyclic Scheduling Scheme

Cyclic scheduling was proposed for a model based on a hierarchical job-flow management scheme [27]. Job-flow scheduling is performed in cycles by separate job batches on the basis of dynamically updated local schedules of computational nodes (Fig. 1). Sets of available slots and their costs (Cj in Fig. 1) determined by resource owners are updated based on the information from local resource managers or job batch processing systems. Thus, during every scheduling cycle two problems have to be solved. First of all, the alternative sets of slots (alternative offers for each batch job) that meet the requirements (resource, time, and cost) should be selected. Each alternative is characterized by the total execution cost, runtime, start time, finish time and other parameters (for example power consumption). Second, a combination of alternatives that would be the efficient or optimal in terms of the whole job batch execution in the current scheduling cycle is chosen (according to the VO policy).

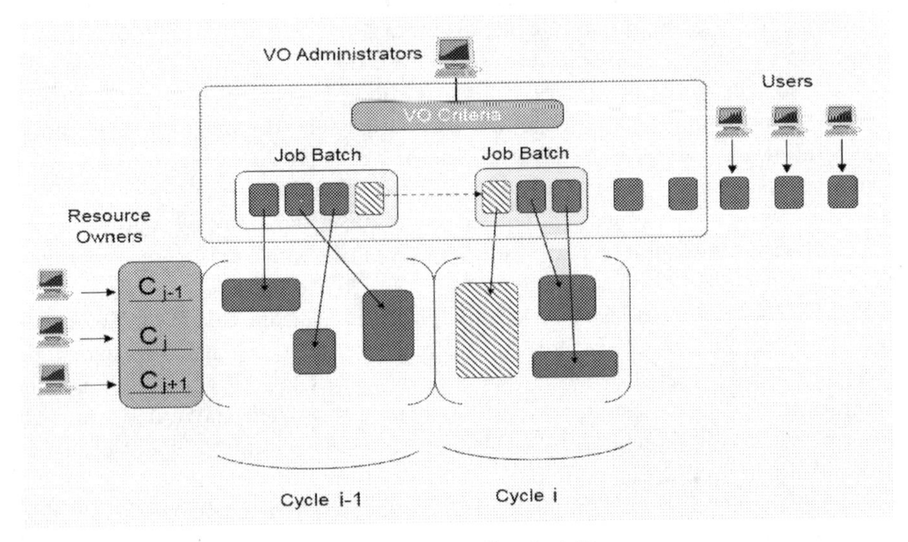

Fig. 1. Job flow cyclic scheduling

Let S_i be the family of appropriate sets of slots for executing job i, $i = 1,...,n$, in the batch, $s_j \in S_i$ be the set of slots that are appropriate by the resource request, the cost $c_i(s_j)$ and the execution time $t_i(s_j)$, $j = 1,...,N, N = \left|\bigcup_{i=1}^{n} S_i\right|$. Denote by S the family of appropriate sets of slots and by $\bar{s} = (s_1,...,s_n)$, $\bar{s} \in S$, the sequence, which we call the combination of slots, for executing the batch of jobs. Let $f_i(s_j)$ be a function determining the efficiency of executing job i in the batch on the set of slots s_j under the admissible expenses specified by the function $g_i(s_j)$. For example, $f_i(s_j) = c_i(s_j)$ is the price of using the set s_j for the time $g_i(s_j) = t_i(s_j)$. The expenses are admissible if $g_i(s_j) \le g_i \le g^*$, where g_i is the level of the total expenses for the execution of a part of jobs from the batch (for example, jobs

$i,i+1,...,n$ or $i,i-1,...,1$) and $g*$ is the restriction for the entire set of jobs (in particular, the restriction on a total time $t*$ of slot occupation or a limitation on a budget $b*$ of the virtual organization).

Formally, the statement of the problem of the optimal choice of a slot combination $\overline{s}=(s_1,...,s_n)$ is as follows:

$$f(\overline{s})=\sum_{i=1}^{n} f_i(s_j)\to \text{extr}\,,\; g_i(s_j)\le g_i \le g*\,,\; g*=\sum_{i=1}^{n} g_i^0(s_j), \tag{1}$$

where $g_i^0(s_j)$ is the resource expense level function of executing the batch.

The recurrences for finding the extremum of the criterion in (1) for the set of slots $s_j\in S_i,\;\; i=\overline{1,n}, j\in\{1,...,N\}$ based on backward recursion are

$$f_i(g_i)=\underset{s_j\in S_i}{\text{extr}}\{f_i(s_j)+f_{i+1}(g_i-g_i(s_j))\},\; g_i(s_j)\le g_i\le g*\,,\; i=\overline{1,n}\,,$$

$$f_{n+1}(g_{n+1})\equiv 0\,,\; g_i=g_{i-1}-g_{i-1}(s_k)\,,\; 1<i\le n\,,\; g_1=g*\,,\; s_k\in S_{i-1}\,, \tag{2}$$

where g_i are the total expenses (utilization time or cost) for using the slots for jobs $i,i+1,...,n$ of this batch.

The optimal expenses are determined from the equation

$$g_i^*(s_j)=\underset{g_i(s_j)\le g_i}{\arg\text{extr}} f_i(g_i)\,,\; i=\overline{1,n}\,. \tag{3}$$

The optimal set of slots $s_i^*\in\{1,...,N\}$ in the scheme (2), (3) is given by the relation

$$s_i^*=\arg\underset{s_j\in S_i}{\text{extr}}\, f_i(g_i*(s_j)),\; i=\overline{1,n}\,. \tag{4}$$

Here (4) represents the solution of the problem (1). An example of a resource expense level function in (1) is $t_i^0(s_j)=[\sum_{s_j} t_i(s_j)/\,l_i]$, where l_i is the number of admissible (alternative) sets of slots for the execution of job i, $[\cdot]$ is the ceiling of $t_i^0(s_j)$. Then the constraint on the total time of slot occupation in the current scheduling cycle can have the form

$$t*=\sum_{i=1}^{n} t_i^0(s_j). \tag{5}$$

Let us consider several problems of practical importance.

1. Maximization of profit of resource owners under restrictions on the total time of slot utilization. Suppose it is required to select a set of slots for executing a batch of n jobs so as to maximize the total cost of resource utilization

$$f_i(t_i)=\max_{s_j\in S_i}\{c_i(s_j)+f_{i+1}(t_i-t_i(s_j))\},\; i=1,...,n\,,\; f_{n+1}(t_{n+1})\equiv 0\,. \tag{6}$$

The restriction on the total time of using slots by all the jobs is given by (5).

2. Minimization of the total completion time of a batch of jobs under a restriction on the budget b^* of the virtual organization:

$$f_i(c_i)=\min_{s_j\in S_i}\{t_i(s_j)+f_{i+1}(c_i-c_i(s_j))\},\ i=1,...,n,\ f_{n+1}(c_{n+1})\equiv 0. \quad (7)$$

3. Minimization of the total cost of executing a batch of n jobs under a restriction on the total time (5) of slot utilization:

$$f_i(t_i)=\min_{s_j\in S_i}\{c_i(s_j)+f_{i+1}(t_i-t_i(s_j))\},\ i=1,...,n,\ f_{n+1}(t_{n+1})\equiv 0. \quad (8)$$

4. Minimization of the idleness of resources under the restriction on the total time of their utilization. On the one hand, the resource owners restrict the time of slot utilization to balance their own (local) and users' job flows. On the other hand, the owners naturally strive to minimize the idleness of resources. Assume that the slot utilization time is bounded by (5). The problem is reduced to finding a set of slots that satisfy this restriction:

$$f_i(t_i)=\max_{s_j\in S_i}\{t_i(s_j)+f_{i+1}(t_i-t_i(s_j))\},\ i=1,...,n,\ f_{n+1}(t_{n+1})\equiv 0. \quad (9)$$

The above functional equations (6)-(9) are concretizations of (2) and are implemented as simulation environment components [27].

Among the major CSS restrictions in terms of an efficient scheduling and resource allocation one may outline the following. First of all, it is not possible to affect execution parameters of an individual job: the search for particular alternatives is performed on the First Fit principle, while choice the optimal combination (4) represents only the interests of VO upon the whole. Thus this approach does not take into account user interests and preferences, and therefore obstructs fair resource sharing. Second, the job batch scheduling is based on a user estimation of the particular job runtime $t_i(s_j)$ (often inaccurate). Thus, in case of estimation incorrectness, the early released resources may become idle reducing the distributed environment utilization level. Third, the job batch scheduling requires allocation of a multiple "nonintersecting" in terms of slots alternatives, and at the same time only one alternative is chosen for each job execution.

Fig. 2 shows a job batch scheduling example consisting of five independent jobs.

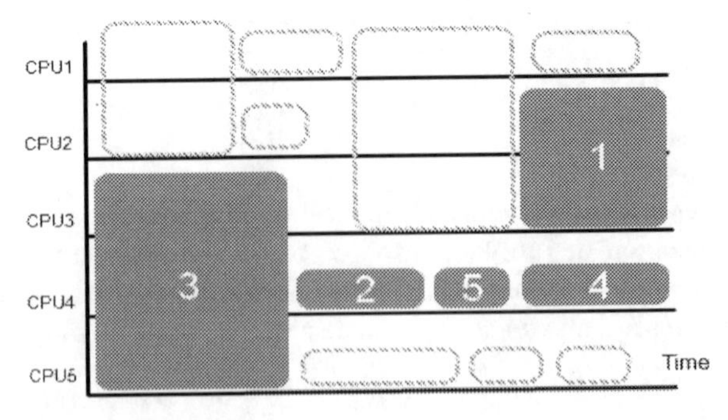

Fig. 2. An example of alternatives allocation for a batch of five jobs

Highlighted rectangles schematically represent all “nonintersecting” in terms of slots alternatives found for the batch on the scheduling cycle in “CPU – Time” space. Filled rectangles represent a combination of the alternatives selected by the metascheduler. Thus, available resources are fragmented, and their utilization level, especially at the beginning of the considered scheduling interval, is relatively low.

The following section is dedicated to the CSS generalization and further development.

3 The Model of Scheduling and Fair Resource Sharing

For the metascheduling concept implementation we put the following requirements for the model of scheduling and fair resource sharing among the VO stakeholders (we name this model as Batch-slicer). First, VO administrators should be able to manage the scheduling process by establishing a job-flow execution policy. Second, VO users should have an opportunity to affect their jobs execution schedule by setting an optimization criterion. Third, resource owners should be able to control utilization level of their computational nodes by specifying their pricing model during the scheduling interval.

Batch-slicer is a generalization of the CSS system described above, and therefore it takes into account the interests of diverse VO stakeholders. In order to satisfy the user preferences a desirable optimization criterion is introduced into the resource request format (Fig. 3).

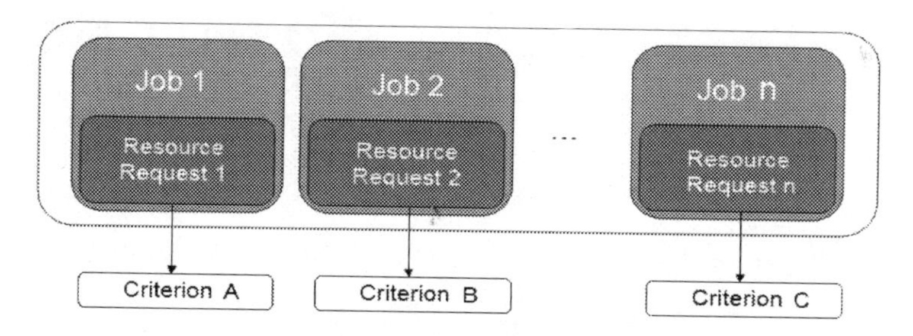

Fig. 3. Users' optimization criteria for jobs execution

Unlike the so-called soft constraints [14] representing the user preferences, the optimization criterion defined in the resource request is considered during the stage of alternatives (slot sets) search. **A**lgorithm searching for **E**xtreme **P**erformance (AEP) described in details in [29] is used to select optimal alternatives under a given criterion. Thus, a set of job execution alternatives is formed by the user preferences according to the individual application optimization criteria. At the same time the optimal alternatives combination choice is carried out in accordance with the criterion which implements VO policy. Resource owners receive an opportunity to manage their own profit and computational nodes utilization by varying local schedules and price establishing during the scheduling cycle.

Another Batch-slicer difference from CSS consists in the job system formation algorithm. Batch-slicer implies a separation of the initial job batch into a set of sub-batches and each sub-batch scheduling at the same given scheduling interval. The idea of "slicing" can be particularly noticeable at a relatively high distributed environment resources utilization level. According to the alternatives search algorithm adopted in CSS [27], the number of execution alternatives for a job batch may be relatively small (up to just a single alternative for every job at a high resource utilization level). Such a small number of alternatives found may affect the optimal slot combination selection (4), and therefore, may reduce overall scheduling efficiency. The job batch "slicing" increases the number of alternatives found for high-priority jobs and diversifies the choice on the slots combination selection (4) stage, and thereby increases the resource sharing efficiency according to VO policy. When separating the original batch to n sub-batches, where n is a total number of jobs in the batch (see subsection 2.3), the algorithm will find the best sets of slots for each job according to the criteria specified in their resource requests. But in this case the efficiency of a whole job batch scheduling is not taken into account. On the other hand, when only a single sub-batch is "picked" from the original job batch the scheduling result will be identical to CSS application.

In view of described modifications, Batch-slicer is schematically shown in Fig. 4: an optimization criterion is specified for each job, and the job batch is separated to the sub-batches during the scheduling cycle.

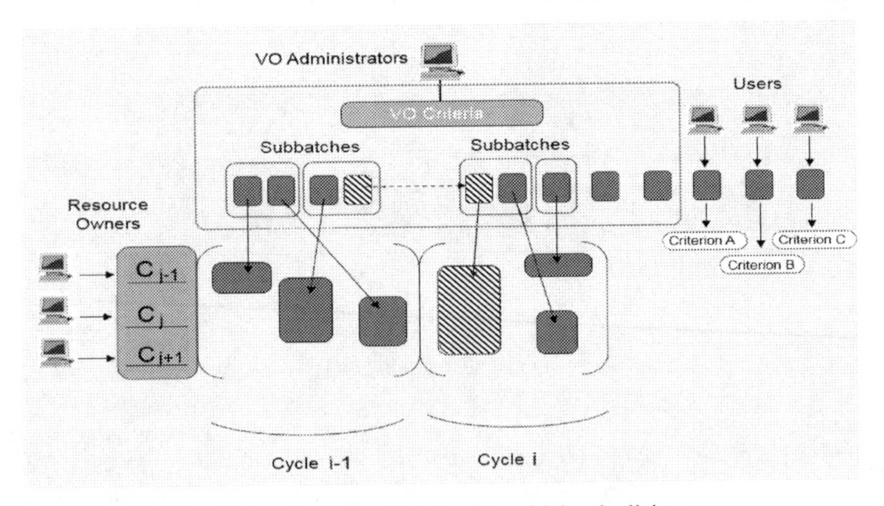

Fig. 4. Job flow cyclic scheduling with batch-slicing

4 Cyclic Scheduling Method Combined with Backfilling

Each of the approaches described above has its advantages and disadvantages. Batch-slicer makes it possible to optimize the job-flow execution according to the VO shareholders preferences on condition that a sufficient number of alternatives was found for the batch jobs during the scheduling cycle. Backfilling responds to early resources releases and performs "on the fly" rescheduling which is very important when a user job runtime estimation is significantly different from the actual job execution time.

We propose a combined approach. During every scheduling cycle a set of high priority jobs, for example the most "expensive" (by total execution cost) or the most critical in terms of required resource (by performance), is allocated from the initial job batch. These jobs should be scheduled before other jobs, probably, without complying the queue discipline. High priority jobs are grouped into a separate sub-batch. The scheduling of this sub-batch is further performed by Batch-slicer based on the preliminary known resources utilization schedule. The scheduling of the rest batch jobs is performed by backfilling with the dynamically updated information about the actual computational nodes utilization. Thus, the cyclic scheduling method combined with backfilling (Batch-slice-Filling - BSF) combines the main advantages of both Batch-slicer and backfilling, namely the optimization of the most time-consuming jobs execution as well as the efficient resource usage, preferential job execution queue order compliance and relatively low response time. The exact number of jobs to select into the first sub-batch to schedule with Batch-slicer and the selection principle may depend on the related resource domain characteristics as well as on the job batch composition and general parameters.

5 Simulation Environment Setup

A series of studies were carried out with the simulation environment [27] in order to investigate the proposed job batch scheduling approaches and to compare them with known scheduling algorithms.

The scheduling environment core consists of the following major components: computational procedures and random variable functions implementation for the environment parameters generation; resource requests and distributed computing environment generation; AEP slot processing; an algorithm for optimal alternatives combination selection; Batch-slicer module; backfilling module; BSF module.

The main features of the simulation environment are as follows.

1. The job-flow and domain heterogeneous resources generation is performed in accordance with the random variables distribution functions with settings specified in the model for the real traces simulation.

2. Initial domain node utilization level is determined by the local tasks number and runtime. The initial CPU node utilization schedule is generated with the hypergeometric distribution.

3. The model supports different pricing mechanisms and the interaction between the VO stakeholders with economic principles.

4. The algorithms for job system formation, alternatives search and the best alternatives combination selection are implemented in the model.

The model components general settings are used for the experiments as follows. A typical scheduling interval length is assumed to be 600 units of time. The number of nodes in the resource domain is equal to 24. The nodes performance level is given as a uniformly distributed random value $p \in [2, 10]$. Thus the resources with the highest performance level (p = 10) are generally able to execute jobs roughly twice as fast as medium performance level nodes (p = 6), while nodes with the lowest performance (p = 2) are three times slower. This configuration provides a sufficient resources diversity level while the difference between the highest and the lowest resource performance levels will not exceed one order within a particular resource domain. Uniform distribution was chosen in the assumption that the CPU node composition is formed by resource selection based on such characteristics as a CPU node type, performance, locations, etc. (hard constraints according to [14]). The node prices are assigned during the pricing stage depending on the node performance level and a random "discount/extra charge" value which is normally distributed. The number of user jobs in each scheduling cycle is assumed to be 20. The jobs budget limit is generated in such a way that the "richest" users can afford to use "expensive" resources with the price formed as a "market value + 60% extra charge", and the "poorest" users have been forced to rely on 60% discounts. These factors prevent the monopoly for the most expensive and, therefore, the high-performance resources.

A special study is a simulation of a complete scheduling cycle for the same job batch independently by proposed and known algorithms. In each experiment, first of all, a job batch presented as a resource requests list is performed, and, second, a resource environment composition with local utilization schedules is generated. Thus the study is based on scheduling results obtained with the same input (job batch) using different scheduling algorithms comparison.

6 Experimental Studies of Fair Resource Sharing

The goal of the investigation is to verify basic concepts of fair resource sharing, i.e. to prove that each VO member has a possibility to affect the process of scheduling according to his preferences (see Section 2).

6.1 Taking into Account VO Users' Preferences

Taking into account VO users' preferences and the analysis of scheduling results for individual jobs is carried out by comparing the proposed Batch-slicer approach with the initial CSS. The latter does not perform optimization during the stage of alternatives selection. Thus, there are two approaches considered in the experiment. First of all, scheduling of the job batch with Batch-slicer where alternatives search is performed based on AEP taking into account the criterion specified in resource request. Second, scheduling of the job batch with CSS where alternatives search is performed by choosing the first fit alternative.

Table 1 shows the results of individual jobs scheduling depending on the optimization criterion specified by the user: job start and finish time, execution time and cost (AEP minimizes the value of the specified criterion).

Table 1. Scheduling results with VO users' preferences

Criterion	N_A	Start time	Execution time	Finish time	Cost
Start time	12.8	171.7	56.1	227.8	1281.1
Execution time	10.6	214.5	***39.3***	253.9	1278.5
Finish time	12.2	***169.6***	45	***205.5***	1283.2
Cost	12.9	262.6	55.5	318	***1098.3***
CSS	*12.1*	222	50.3	272.3	1248.4

The choice of one of the four optimization criteria is made randomly with uniform distribution at the stage of job batch generation. Uniform distribution is used because no prevalent optimization criterion can be chosen. The last row of Table 1 shows the results of scheduling of the same job batch with initial CSS without optimization at the stage of alternatives search. Simulation of 5000 individual scheduling cycles was conducted. As can be seen from Table 1, best values against start and finish time criteria as well as by execution time and cost (the minimal values are marked in bold) are achieved by the jobs for which the corresponding optimization criterion is specified ("Criterion" column). The only exception is finish time minimization approach: the jobs for which this optimization criterion was specified show on average not only the minimal finish time, but also the minimal start time. On average the use of an optimization criterion in Batch-slicer, in comparison with CSS, when executing individual jobs, allows reducing job start and finish time by more than 23%, reducing execution time by 21% and reducing execution cost by 12%. Average number of execution alternatives (N_A in Table 1) found for the jobs during one scheduling cycle almost does not depend on the chosen optimization criterion. Average number of jobs per each group having the same optimization criterion equals 5 on average. This fits the use of uniform distribution when choosing one of the four optimization criteria for each of the 20 batch jobs.

The individual jobs scheduling results show that users can affect the execution of their own jobs by specifying an optimization criterion. This is achieved due to the fact that in the presence of different requirements to efficiency of job execution resources are allocated among the jobs unevenly, depending on the criterion used in selection. Note that in the initial CSS at the stage of alternatives search all the resources are allocated among the jobs uniformly.

6.2 Optimization of Job Batch Execution in VO

The next experiment is dedicated to comparing the scheduling results when slicing the initial job batch in Batch-slicer into different number of sub-batches and at different levels of environment utilization. The experiment allows estimating the efficiency of scheduling in different modes with different input data. Modes comparison was performed on the basis of job batch allocation results on full scheduling cycle including initial environment generation, composition of batches and sub-batches and then their sequential scheduling.

When choosing the optimal execution alternatives combination the average job execution time T_{CPU} minimization task was being solved. Total slot utilization time for an alternative is determined as the sum of slot lengths being part of the composed "window". Fig. 5 shows the value of T_{CPU} depending on the number of sub-batches $k \in \{1,2,3,5,6,10,20\}$ into which the initial batch is sliced and the level of environment utilization. When performing the series of experiments the environment utilization level is determined by the relative average number of failures Y – scheduling cycles in the course of which the execution schedule for all the batch jobs was not found. The experiments were conducted under high ($Y = 0.3$), medium ($Y = 0.03$) and low utilization levels ($Y < 0.0002$). Thus the number of failures in the conducted series of experiments differs at the minimum by the order of magnitude of one.

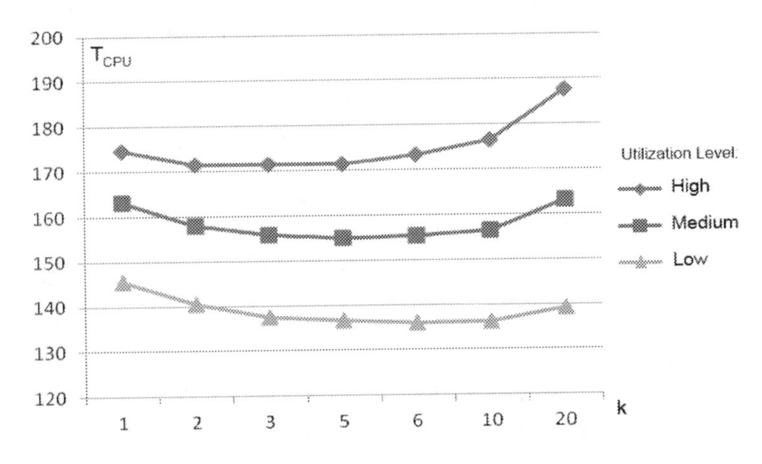

Fig. 5. Average batch jobs execution time T_{CPU} depending on the number of subbathes k

As a result of the job batch scheduling experiment the following patterns were revealed. An increase of composed sub-batches number causes an increase of alternatives number for execution an individual job, a decrease of total job execution cost, and an increase of relative number of failures Y. When increasing the level of available resources the number of alternatives for an individual job execution increases, the relative number of failures Y decreases, and the total cost of job batch execution decreases. The experiment results show that slicing of the initial batch into sub-batches and their sequential independent scheduling allows increasing the number of execution alternatives for the batch jobs and performing more efficient execution schedules. So, at a high execution environment utilization rate the best efficiency and the least number of failures is provided by slicing into fewer sub-batches. On the other hand, at a low utilization of available resources it is advantageous to slice into a greater number of sub-batches up to scheduling the jobs individually (see Fig. 5). Another advantage of Batch-slicer in comparison with CSS is decreasing of total execution cost as the level of resource utilization becomes lower. CSS tries to use the entire admissible budget b^* for the job batch execution by choosing the corresponding set of alternatives. At the same time when scheduling sub-batches with a small number of jobs the choice is often confined to a few alternatives whose cost is not necessarily close to the admissible budget limitation.

Thus, Batch-slicer allows not only taking into account VO administrators' preferences (by optimizing at the alternatives set selection stage, like in the initial CSS), but can also provide a better value of the target criterion in comparison with CSS by slicing into sub-batches (the least value of the target criterion – job batch total execution time – was achieved when slicing the job batch into 5 sub-batches with four jobs in each of them).

6.3 Taking into Account VO Resource Owners' Preferences

Table 2 shows the scheduling results with Batch-slicer from resource owners' point of view by the example of a single CPU node characteristics depending on the unit cost c, specified for the use of scheduling interval $T = 600$: L_c – total slot utilization time in the scheduling interval; U – relative resource utilization average value in the scheduling interval; S – average profit made by the resource owner, and Y – relative number of scheduling failures.

Table 2. Scheduling results with VO resource owners' preferences

c	L_c	U	S	Y
2	256.6	0.44	527.1	0
4	234.9	0.39	939.6	0.001
6	185.4	0.31	1112.3	0.013
8	109.8	0.18	878.7	0.024
10	71	0.12	710.3	0.025

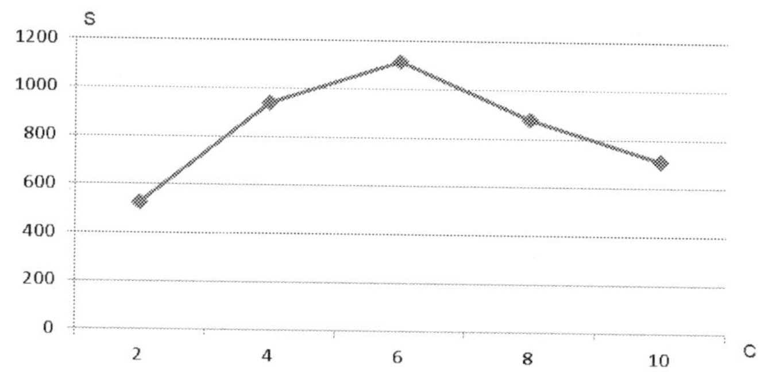

Fig. 6. A resource owner's profit S depending on the proposed price c

As can be seen from Table 2, resource owners are able to control their profit S and the computational node utilization level U in the scheduling interval T by proposing the unit cost c of using their node. Profit extremum is achieved when proposing the cost close to the "average market cost", i.e. the average cost for a resource with similar performance, proposed by other resource owners. The profit value received by a resource owner for providing a single computational node is illustrated graphically in Fig. 6.

7 Experimental Studies of Resource Use Efficiency in the Cyclic Scheme

7.1 Studies of Combined Scheduling Method BSF

The efficiency of scheduling with BSF combined approach can be considered from two viewpoints at the same time: on the one hand, from the viewpoint of criterion value optimization in the specific VO, job batch total execution time (7), for instance, and, on the other hand, from the environment utilization level and batch job start time minimization viewpoint.

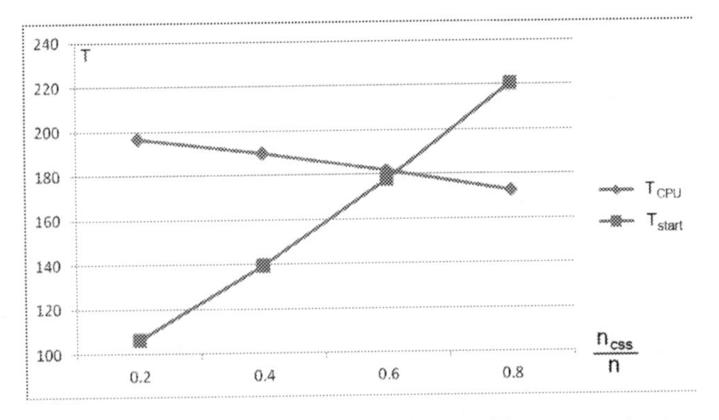

Fig. 7. Average job execution T_{CPU} and start T_{start} time in BSF

Fig. 7 shows batch job average execution time T_{CPU} and average start time T_{start} depending on the ratio according to which slicing into sub-batches was made: n_{CSC} – the number of jobs in the first sub-batch, scheduled with CSS, n – total number of jobs in the batch. Slicing into sub-batches was made on basis of a priority – the order of jobs in the batch, without taking into account the characteristics of the jobs themselves. The scheduling results presented in Fig. 7 are obtained based on simulation of 5000 independent scheduling cycles at a medium level of resource utilization according to the settings described in section 5. As seen from Fig. 7, if a major part of the job is scheduled with Batch-slicer then a better value of the target VO scheduling criterion – execution time T_{CPU} – is achieved, but average job start time T_{start} is delayed. And on the contrary, if a major part of the job is scheduled with backfilling, then average start time approaches the beginning of the scheduling interval but the value of the target optimization criterion deteriorates. Particular emphasis should be placed on the cross point of graphs in Fig. 7. Its presence given that the graphs are monotone suggests the possibility of reaching a compromise between average start time and the value of the VO target optimization criterion. So, BSF shows "compromise" values of the discussed characteristics compared to BS and backfilling.

7.2 Experimental Studies of Consistency of Schedules Based on Job Execution Time Estimate

Let us consider scheduling efficiency studies results and consistency of schedules performed with Batch-slicer and CSS and based on job execution time estimate which is specified in the resource request. Batch-slicer and CSS form preliminary job batch execution schedules in the scheduling interval without taking into account the situations in which real job execution time is less than the time specified by the user. Early job completion, untimely resource release and idleness may negatively affect the efficiency of job batch execution against the criteria specified by VO stakeholders and make the schedule inconsistent. On the other hand, backfilling conducts scheduling on basis of dynamically updated information on job execution status and computational node utilization. Thanks to this it can provide the efficient job flow execution. A simulation was conducted to study and to compare the efficiency of schedules performed with CSS, Batch-slicer and backfilling. In the simulation real job execution time differed considerably from resource advanced reservation time. Real job execution time was specified as a random variable uniformly distributed in the interval $[0.2*T_{res}, T_{res}]$, where T_{res} – time reserved for job execution. Uniform distribution is chosen as it is almost impossible to predict real job execution time on the specified resources. Thus, at worst real execution time could differ from user estimate by 5 times.

Table 3 contains the average job execution time values (the target optimization criterion) and average job start time obtained: 1) at the stage of preliminary scheduling based on job execution time estimate T_{res} ("Scheduled" column); 2) as the result of execution simulation of the composed schedule taking into account real job execution time on the chosen resources ("Real" column).

Table 3. Real and scheduled job execution time

Algorithm	Execution time		Average job start time	
	Scheduled	Real	Scheduled	Real
Backfilling	187.7	115.1	69	37.3
CSS	150.1	90.4	281.2	281.2
Batch-slicer	138.6	83.5	223.8	223.8
Advantage of Batch-slicer over backfilling	26.2%	27.5%	-69%	-83%

It can be seen from Table 3 that even if the difference between resource reservation time and real job execution time is significant the advantage of Batch-slicer over backfilling against the VO target optimization criterion not only remains but increases. That is because backfilling does not optimize against criteria different from start time and a more compact job location (real start time of jobs scheduled with backfilling is reduced by 46% on average) uses almost all the available resources including those less advantageous against the target criterion.

Thus, results of the experiment show that preliminary schedules formed in the beginning of the scheduling cycle are consistent against the criteria determined in VO in the case when real execution time differs significantly from resource reservation time. Note that additional advantage can be achieved by rescheduling taking into account the information about computational nodes' current utilization.

8 Conclusions and Future Work

In this work, we address metascheduling strategies with different target criteria and based on scheduling and fair resource sharing model taking into account all VO stakeholders' preferences on the basis of economic principles. A solution to the problem of fair resource sharing among VO stakeholders is proposed.

The advantage over initial CSS when scheduling the job flow reaches 7% and in the terms of single job execution it reaches 25% at a medium level of environment utilization. Resource owners can vary the resource provision unit cost (by offering discounts for instance) to maximize the profit or to achieve the necessary resource utilization level. Based on union of CSS and backfilling a combined approach BSF is proposed. The approach shows compromise results compared to Batch-slicer and backfilling, namely it allows utilizing the available resources efficiently (by means of backfilling) when efficiently executing a part of jobs in VO (by means of optimization in Batch-slicer). The consistency of scheduling made with Batch-slicer when real job execution time is significantly different from user's estimate is shown.

Further research will be related to a more precise investigation of dividing the job flow into sub-batches depending on the jobs characteristics and computing environment parameters as well as to studying the mechanism of rescheduling based on the information about computational nodes current utilization.

Acknowledgements. This work was partially supported by the Council on Grants of the President of the Russian Federation for State Support of Leading Scientific Schools (SS-362.2014.9), the Russian Foundation for Basic Research (grant no. 12-07-00042), and by the Federal Target Program "Research and scientific-pedagogical cadres of innovative Russia" (state contract no. 16.740.11.0516).

References

1. Garg, S.K., Buyya, R., Siegel, H.J.: Scheduling Parallel Applications on Utility Grids: Time and Cost Trade-off Management. In: 32nd Australasian Computer Science Conference (ACSC 2009), Wellington, New Zealand, pp. 151–159 (2009)

2. Degabriele, J.P., Pym, D.: Economic Aspects of a Utility Computing Service, Trusted Systems Laboratory HP Laboratories Bristol HPL-2007-101. Technical Report, pp. 1–23 (July 3, 2007)
3. Garg, S.K., Yeo, C.S., Anandasivam, A., Buyya, R.: Environment-conscious Scheduling of HPC Applications on Distributed Cloud-oriented Data Centers. J. Parallel and Distributed Computing 71(6), 732–749 (2011)
4. Tesauro, G., Bredin, J.L.: Strategic Sequential Bidding in Auctions Using Dynamic Programming. In: 1st International Joint Conference on Autonomous Agents and Multiagent Systems, Part 2, pp. 591–598. ACM, New York (2002)
5. Thain, D., Tannenbaum, T., Livny, M.: Distributed Computing in Practice: the Condor Experience. J. Concurrency and Computation: Practice and Experience 17(2-4), 323–356 (2004)
6. Berman, F.: High-performance Schedulers. In: Foster, I., Kesselman, C. (eds.) The Grid: Blueprint for a New Computing Infrastructure, pp. 279–309. Morgan Kaufmann, San Francisco (1999)
7. Yang, Y., Raadt, K., Casanova, H.: Multiround Algorithms for Scheduling Divisible Loads. IEEE Trans. Parallel and Distributed Systems 16(8), 1092–1102 (2005)
8. Natrajan, A., Humphrey, M.A., Grimshaw, A.S.: Grid Resource Management in Legion. In: Nabrzyski, J., Schopf, J.M., Weglarz, J. (eds.) Grid Resource Management. State of the Art and Future Trends, pp. 145–160. Kluwer Academic Publishers, Boston (2003)
9. Beiriger, J., Johnson, W., Bivens, H.: Constructing the ASCI Grid. In: 9th IEEE Symposium on High Performance Distributed Computing, pp. 193–200. IEEE Press, New York (2000)
10. Frey, J., Foster, I., Livny, M.: Condor-G: A Computation Management Agent for Multi-institutional Grids. In: 10th International Symposium on High-Performance Distributed Computing, pp. 55–66. IEEE Press, New York (2001)
11. Abramson, D., Giddy, J., Kotler, L.: High Performance Parametric Modeling with Nimrod/G: Killer Application for the Global Grid? In: International Parallel and Distributed Processing Symposium, pp. 520–528. IEEE Press, New York (2000)
12. Foster, I., Kesselman, C., Tuecke, S.: The Anatomy of the Grid: Enabling Scalable Virtual Organizations. Int. J. of High Performance Computing Applications 15(3), 200–222 (2001)
13. Ranganathan, K., Foster, I.: Decoupling Computation and Data Scheduling in Distributed Data-intensive Applications. In: 11th IEEE International Symposium on High Performance Distributed Computing, pp. 376–381. IEEE Press, New York (2002)
14. Kurowski, K., Nabrzyski, J., Oleksiak, A., Weglarz, J.: Multicriteria Aspects of Grid Resource Management. In: Nabrzyski, J., Schopf, J.M., Weglarz, J. (eds.) Grid Resource Management. State of the art and Future Trends, pp. 271–293. Kluwer Academic Publishers, Boston (2003)
15. Garg, S.K., Konugurthi, P., Buyya, R.: A Linear Programming-driven Genetic Algorithm for Meta-scheduling on Utility Grids. J. Par., Emergent and Distr. Systems 26, 493–517 (2011)
16. Buyya, R., Abramson, D., Giddy, J.: Economic Models for Resource Management and Scheduling in Grid Computing. J. Concurrency and Computation 14(5), 1507–1542 (2002)
17. Ernemann, C., Hamscher, V., Yahyapour, R.: Economic Scheduling in Grid Computing. In: Feitelson, D.G., Rudolph, L., Schwiegelshohn, U. (eds.) JSSPP 2002. LNCS, vol. 2537, pp. 128–152. Springer, Heidelberg (2002)
18. Lee, Y.C., Wang, C., Zomaya, A.Y., Zhou, B.B.: Profit-driven Scheduling for Cloud Services with Data Access Awareness. J. Par. and Distr. Computing 72(4), 591–602 (2012)

19. Toporkov, V.V.: Job and Application-Level Scheduling in Distributed Computing. Ubiquitous Computing and Communication J. Applied Computing 4(3), 559–570 (2009)
20. Toporkov, V.V., Toporkova, A., Tselishchev, A., Yemelyanov, D.: Job and Application-Level Scheduling: an Integrated Approach for Achieving Quality of Service in Distributed Computing. In: 4th International Conference on Dependability of Computer Systems, pp. 202–209. IEEE CS Press, Los Alamitos (2009)
21. Toporkov, V.: Application-Level and Job-Flow Scheduling: an Approach for Achieving Quality of Service in Distributed Computing. In: Malyshkin, V. (ed.) PaCT 2009. LNCS, vol. 5698, pp. 350–359. Springer, Heidelberg (2009)
22. Aida, K., Casanova, H.: Scheduling Mixed-parallel Applications with Advance Reservations. In: 17th IEEE Int. Symposium on HPDC, pp. 65–74. IEEE CS Press, New York (2008)
23. Ando, S., Aida, K.: Evaluation of Scheduling Algorithms for Advance Reservations. Information Processing Society of Japan SIG Notes HPC-113, 37–42 (2007)
24. Elmroth, E., Tordsson, J.: A Standards-based Grid Resource Brokering Service Supporting Advance Reservations, Coallocation and Cross-Grid Interoperability. J. of Concurrency and Computation 25(18), 2298–2335 (2009)
25. Toporkov, V., Toporkova, A., Bobchenkov, A., Yemelyanov, D.: Resource Selection Algorithms for Economic Scheduling in Distributed Systems. Procedia Computer Science 4, 2267–2276 (2011)
26. Bailey Lee, C., Schwartzman, Y., Hardy, J., Snavely, A.: Are User Runtime Estimates Inherently Inaccurate? In: Feitelson, D.G., Rudolph, L., Schwiegelshohn, U. (eds.) JSSPP 2004. LNCS, vol. 3277, pp. 253–263. Springer, Heidelberg (2005)
27. Toporkov, V., Tselishchev, A., Yemelyanov, D., Bobchenkov, A.: Dependable Strategies for Job-flows Dispatching and Scheduling in Virtual Organizations of Distributed Computing Environments. In: Zamojski, W., Mazurkiewicz, J., Sugier, J., Walkowiak, T., Kacprzyk, J. (eds.) Complex Systems and Dependability. AISC, vol. 170, pp. 289–304. Springer, Heidelberg (2012)
28. Moab Adaptive Computing Suite, `http://www.adaptivecomputing.com/products/moab-adaptive-computing-suite.php`
29. Toporkov, V., Toporkova, A., Tselishchev, A., Yemelyanov, D.: Slot Selection Algorithms for Economic Scheduling in Distributed Computing with High QoS Rates. In: Zamojski, W., Mazurkiewicz, J., Sugier, J., Walkowiak, T., Kacprzyk, J. (eds.) New Results in Dependability & Comput. Syst. AISC, vol. 224, pp. 459–468. Springer, Heidelberg (2013)
30. Azzedin, F., Maheswaran, M., Arnason, N.: A Synchronous Co-allocation Mechanism for Grid Computing Systems. Cluster Computing 7, 39–49 (2004)
31. Castillo, C., Rouskas, G.N., Harfoush, K.: Resource Co-allocation for Large-scale Distributed Environments. In: 18th ACM International Symposium on High Performance Distributed Compuing, pp. 137–150. ACM, New York (2009)
32. Takefusa, A., Nakada, H., Kudoh, T., Tanaka, Y.: An Advance Reservation-based Co-allocation Algorithm for Distributed Computers and Network Bandwidth on QoS-guaranteed Grids. In: Frachtenberg, E., Schwiegelshohn, U. (eds.) JSSPP 2010. LNCS, vol. 6253, pp. 16–34. Springer, Heidelberg (2010)
33. Blanco, H., Guirado, F., Lérida, J.L., Albornoz, V.M.: MIP Model Scheduling for Multi-clusters. In: Caragiannis, I., Alexander, M., Badia, R.M., Cannataro, M., Costan, A., Danelutto, M., Desprez, F., Krammer, B., Sahuquillo, J., Scott, S.L., Weidendorfer, J. (eds.) Euro-Par Workshops 2012. LNCS, vol. 7640, pp. 196–206. Springer, Heidelberg (2013)

Improvement of Dependability of Complex Web Based Systems by Service Reconfiguration

Tomasz Walkowiak and Dariusz Caban

Wrocław University of Technology, Wybrzeże Wyspiańskiego 27, 50-320 Wrocław, Poland
{tomasz.walkowiak,dariusz.caban}@pwr.wroc.pl

Abstract. Web based information systems are exposed to various dependability issues during their lifetime (originating in the hardware, in the software or stemming from security vulnerabilities). We present an approach to use reconfiguration to circumvent these issues. The presentation is focused on the potential threads, measuring the dependability risks, constructing of optimal reconfiguration strategies and assessing their impact on the over-all dependability. The proposed technique involves construction of the reconfiguration graph, assessment of the steady state probabilities of web system dependability states and choosing the optimal strategy from among the admissible ones.

Keywords: Web based systems, dependability, security, reconfiguration graph, optimization.

1 Introduction

Whenever a fault manifests itself in a Web based system, whether it is a hardware failure, a software error or a security attack, the administrator is faced with the difficult problem to maintain the continuity of business services. Isolation of the affected hardware and software is usually the first reaction (to prevent propagation of the problem to yet unaffected parts of the system). It then follows that the most important services have to be moved from the affected hosts/servers to those that are still operational. This redeployment of services [3, 9] is further called system reconfiguration.

Reconfiguration is realized in critical time constraints, there is risk that the problem may escalate due to untimely or improper administrative decisions. To prevent this, a reconfiguration strategy should be planned beforehand. The problem of assessing the dependability consequences of such a reconfiguration strategy is addressed hereafter.

Since the reconfiguration strategy is planned in advance, as a list of contingency actions to be taken in case of the various foreseen issues, it is desirable to optimize it, so the Web based system is minimally affected. A number of optimization tasks are formulated to achieve this.

W. Zamojski and J. Sugier (eds.), *Dependability Problems of Complex Information Systems*,
Advances in Intelligent Systems and Computing 307, DOI: 10.1007/978-3-319-08964-5_9

2 Dependability of Web Based Systems

The most popular definition of dependability was proposed by A. Avizienis, J.C. Laprie and B. Randell [1]: it is defined as the capability of systems to deliver service that can justifiably be trusted. The following aspects of this definition are of particular interest:

- It relates dependability with the functionality of systems, i.e. with their ability to provide the functionality in presence of faults.
- It relates dependability with justifiable trust, not specifically probability, allowing approaches which are not based on stochastic processes.

A closely related aspect of dependability, even though not expressed directly in the definition, is the way that faults are defined in this approach, i.e. the introduction of the trichotomy of fault – error – failure:

- *Fault* relates to the fact that some system component may be inoperational or may be operating incorrectly. A fault may exist in the system from the beginning of its life cycle (design or production fault, software bugs) or it may occur during its exploitation (natural wear, incidental stresses, hardware/software replacements or upgrades, human errors, security breaches, etc.).
- *Error* relates to the system operation. A fault may be dormant in the system for any extend of time. When the system makes use of a faulty component during its operation, then the corresponding function is not realized correctly. Then, an error is said to occur.
- *Failure* relates to the results of system operation. A failure is said to occur if the results of an error occurrence manifest themselves in the system not producing output or producing incorrect output.

The concept of dependability was introduced to unify the concepts of systems reliability and software reliability. Systems reliability was introduced to the engineering community to explain the phenomena occurring in complex systems (as reported by R. E. Barlow [2]). It was observed that the lifespan of a system was often much shorter than expected on the basis of the quality of the components being used. To understand this and improve the predictions, the reliability was defined as the probability of a device performing its purpose adequately for the period of time intended under operating conditions encountered. The reliability of a system is related to the reliability of its components and its reliability structure.

The definition of dependability is very similar to the above statement, but with significant differences. These differences in the approach address the problems encountered when trying to define the reliability as applying to computer systems, complex fault tolerant digital circuits and, especially, to software. In all these cases the classical definition, connecting reliability to the system structure, cannot be applied. The reliability structure varies depending on the considered system functions.

Software is usually regarded as a system component that is not prone to degrade (acquire faults during exploitation). The term "software reliability" was introduced to capture the similarity of software to hardware operation. In case of software, the faults do not occur during exploitation, they are dormant in the program from its development or are introduced when patching or upgrading it. The faults are not random. Instead, when a program is running it activates different parts of the code in a pseudo-random fashion. When a fault is activated, the software fails to operate correctly. Thus, even though faults are not random, the occurrence of software errors is random.

Software reliability is defined as the probability of error-free software operation for a specified period of time in a specified environment (see [7]). Software reliability growth models relate this probability to the number of faults dormant in the system and the exploitation time.

Even though this definition is very similar to the definition of reliability, the underlying mechanisms of failure are completely different. In reliability theory a component fault can be either masked or cause system failure when it occurs. Software reliability introduces the concept of systems being operational, regardless faulty components that are not masked. The problem is in the visibility of faults. In the definition of dependability, this is dealt with by distinguishing faults, errors and failure.

Web based systems are a combination of failure prone hardware and software. Thus, it is appropriate to use the dependability approach as opposed to systems reliability or software reliability. Furthermore, the concept can also encompass security issues, such as vulnerability exploits, malware proliferation and denial-of-service.

2.1 Faults Taxonomy

As already mentioned, when considering dependability of a Web based system, it is necessary to analyse a very diverse set of faults. It encompasses hardware faults, errors in the software, security vulnerabilities. A taxonomy based on the primary cause of faults is feasible, but it not very useful for these considerations.

The most suitable for the proposed analysis is the classification of faults that is based on the effect it has on the Web system. Particularly, the classification considers the suitability of service relocation as a remedy to the fault.

It should be stressed that the occurrence of a fault may escape detection for some time. This may be the case in all the considered classes of hardware/software faults. It is almost a rule in case of security incidents. In all these cases the incident containment and recovery procedures can be applied only after detection. This also applies to the proposed relocation techniques. For this reason the proposed taxonomy of faults, as described in Fig. 1, is addressed to the detected faults only. Undetected faults can proliferate through the system, eventually causing detected propagation faults, data inconsistencies in the system, and in some cases corrupting some hosts.

In the considered approach, the hosts and communication channels are the basic components of the system. Thus, all the faults are attributed to them (and not to particular hardware or software components). It should also be noted that the communication faults are usually handled at the infrastructure level (by retransmission, error correction techniques, rerouting, etc.). They are rarely allowed to propagate to the system view as discussed in this paper. Thus, even though they are indicated in the taxonomy, we will not consider them as the potential events initiating relocation.

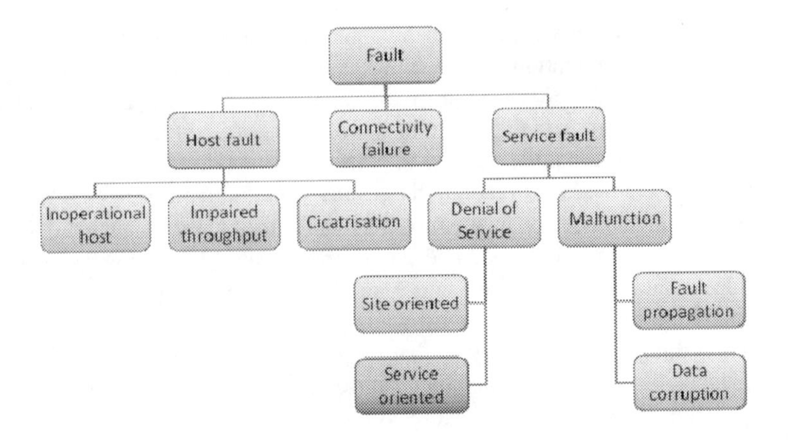

Fig. 1. A classification of system faults reflecting their impact on reconfiguration

The faults can either affect a host or only a service running on it. We distinguish the following classes of faults that affect the host:

Host crash – the host cannot process services that are located on it, these in turn do not produce any responses to queries from the services located on other hosts.
Performance fault – the host can operate, but it cannot provide the full computational resources, causing some services to fail or increasing their response time above the acceptable limits.
Host infection – caused by the proliferation of software errors, effects of transient malfunctions, exploitation of vulnerabilities, malware propagation. The operation of services located on the host becomes unpredictable and potentially dangerous to services at other nodes (service corruption fault). Due to the potential damage that the host may cause, it is usually isolated from the system. This is equivalent to a crash fault with potential service corruption.

The faults that affect a single service can be classified on the basis of their aftereffects as:
Inaccessible service – the service component becomes incapable of responding to requests, due to exploitation of vulnerabilities or a DOS attack. This fault can be location dependent (**location locked fault**), in which case relocation may be a fast and effective remedy. On the other hand, it may be **service locked**, in which case relocation will be ineffective and potentially dangerous to the new location. Relocation should never be applied in this case.
Corrupted service – the service commences to produce incorrect or inconsistent responses due to software errors or vulnerabilities. Usually, this is a propagated fault that can be simply eliminated by restarting the affected software. This type of fault does not need relocation, though relocation will be effective (since it ensures software restart). It should be noted, though, that the effects of a corrupted service propagate to other service components, possibly locating on other hosts. These may also need recovery.

Data inconsistency – propagating errors and malware may cause more persistent effects, by corrupting the system database. This type of faults can be very costly to recover. Technically, though, they are also remedied by service restart from the last valid backup point.

It should be noted that all the faults may lead to system failure if left unhandled. Service relocation may preserve the system functionality, though in some cases it might be an over-reaction. In case of service oriented DOS attacks, relocation is insufficient, requiring additional handling. Otherwise, it might escalate the problem by increasing the extent of penetration.

2.2 Dependability Measures

Dependability is assessed based on the concept of "justifiable trustworthiness" [1]. This trust can be measured using probability. Thus, all the measures used in reliability theory can be also applied to dependability. Since the Web based systems are clearly renewable, so the standard measure of availability may be used.

The availability function $A(t)$ is defined as the probability that the system is operational (provides correct responses) at a specific time t. In stationary conditions, most interesting from the practical point of view, the function is time invariant, characterized by a constant coefficient, denoted as A. The asymptotic property of the steady-state availability A provides a simple formula for assessing it [2]:

$$A = \lim_{t \to \infty} \frac{t_{up}}{t} \tag{1}$$

based on the system total accumulated uptime t_{up}.

In a number of cases it is questionable to assume the probabilistic nature of faults occurrence, especially if they are related with human actions. It is then difficult to uphold the probabilistic interpretation of the observed ratio of uptime to the total time of running the system. Still, the ratio (1) can be used as an independent dependability metric.

Dependability is an integrative concept that encompasses a number of different aspects of system operation in the presence of faults [1]:

- availability (readiness for correct service),
- reliability (continuity of correct service),
- safety (absence of catastrophic consequences),
- confidentiality (absence of unauthorized disclosure of information),
- integrity (absence of improper system state alterations),
- maintainability (ability to undergo repairs and modifications).

Any of the considered faults can affect all of these properties of the Web based systems. Not surprisingly, there are many dependability metrics in use, addressing specific subsets of these properties. This is overviewed in [5]. Many of the metrics are of marginal interest in our considerations, since they are not directly affected by reconfiguration. For this reason, we will further focus on the metrics directly connected with availability and reliability.

Availability coefficient does not reflect the comfort of using the services by the end-users, especially connected with operation in the degraded state. In these situations, the system is operating at 1..L levels of degradation. The quality of service is different at each level of degradation. For each level, a different coefficient of availability A_l can be considered. This coefficient represents the steady-state probability that the system is operating at the l-th level of degradation. It can be determined by a modified equation (1), where total uptime is replaced with accumulated time of operation at level l.

In situations, where there is no single value for availability, it is necessary to use a modified measure, combining the various coefficients. This measure of the overall quality of service Q is determined as:

$$Q = \sum_{l=1}^{L} q_l \cdot A_l \tag{2}$$

The equation assumes that operation at each level of degradation is characterized by a value related to its quality of service q_l . The choice of this measure is not trivial; there are various approaches in use. It can be arbitrarily attributed to each degradation level (e.g. the degradation level can be used directly). More often, it is a measure of system performance. The short discussion of the quality measures used in case web based systems is presented in Section 4.1.

A commonly used, very simple and intuitive approach to dependability assessment is based on identifying the single (and multiple) points of failure in the system. This is in fact a metric of system resilience, i.e. its ability to deliver services after an error occurs. We propose to use two such measures: *SPF* (single points of failure) and *MPF* (multiple points of failure). *SPF* is the number of single faults that cause the system to become inoperational; if *SPF* = 0, then *MPF* is evaluated as the smallest number of faults that must occur for the system to fail.

2.3 Dependability State-Transition Graph

Faults occur in the system randomly, usually with a predictable distribution. Then the system for some time operates in a degraded state or becomes inoperational in a failure state, until maintenance procedures restore it to full operability. The purpose of stochastic analysis is to assess the dependability of the system, especially to assess its improvement when a reconfiguration policy is implemented.

The analysis is based on a stochastic state-transition process, in which the states are described by the vectors of operability states of all the hosts in the system. Assuming that the faults can either have no effect on specific hosts or can cause them to become fully unavailable, the system state is defined as the vector of the up-down states of the hosts. The transitions between states are caused by incident occurrence and by renewal.

The state-transition model can be analyzed using a number of approaches: as a Marcov chain, using semi-Markov processes, using Monte Carlo simulation. The applicability of each method depends on the assumptions that can be made regarding faults occurrence. In case of the Marcov approach, it is necessary to assume that both the faults and renewals occur with constant intensities (i.e. exponential distribution). This approach is used hereafter in the presented case study. A significantly more general approach is possible using simulation, which was investigated by us in [10, 11].

Whatever the method used for analyzing the S-T model, the results are the probabilities that the system is in a specific dependability state at a time instance. This is denoted as P_s, where $s \in \mathbf{S}$ are the possible states.

All the dependability states are classified into two categories: states where the system is still operational $\mathbf{S}_{up}$, and those where it is down $\mathbf{S}_{down}$. If the system is not reconfigured and it does not have any other mechanisms of fault tolerance, then only one of its states (all hosts being up) is in the $\mathbf{S}_{up}$ set. Reconfiguration extends the set of up states to all those situations where there is a configuration in which the system preserves its functionality. Then, the system availability is obtained as:

$$A = \sum_{s \in \mathbf{S}_{up}} P_s \,. \tag{3}$$

When considering the quality of a Web based system that can operate in degraded states, then each dependability state s has to be classified to one of L+1 levels of degradation, not just as up or down. The quality coefficient q_s is determined on the basis of this classification. Then, the system quality is obtained as:

$$Q = \sum_{s \in \mathbf{S}_{up}} q_s \cdot P_s \,. \tag{4}$$

3 Reconfiguration of Web Services

One of the most promising techniques to increase dependability is based on utilizing the functional redundancy of a system. At network level this is routinely achieved by dynamic routing and load balancing. At system level this can be improved by introducing reconfiguration of services when failures occur [6, 8]. This chapter addresses the issues of reconfiguring Web based systems. Reconfiguration is used to improve the dependability of these systems, invoked as a reaction to a fault occurrence. Its aim is to recover the system functionality, fully or partially, while the consequences of the fault occurrence still prevent normal operation.

3.1 Deployment of Services

The Web based systems provide some business service(s), useful to the end-users as a result of interaction between communicating component services, which are transparent to the end-user. In Fig. 2, the system is represented by the interacting service components, which are deployed on various hosts (networked computer nodes). The services make use of the hosts to provide the required processing capabilities, and of the network resources to ensure visibility and data exchange.

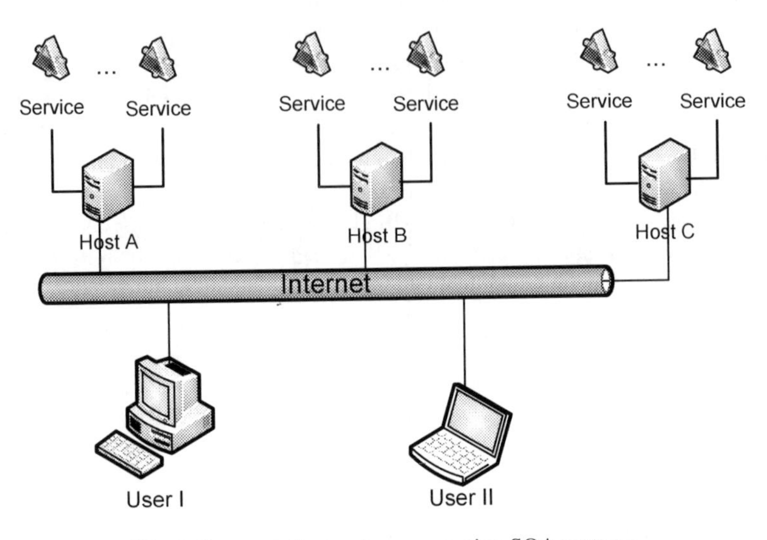

Fig. 2. System infrastructure supporting SOA systems

Each host is characterized by its computing resources: processing power, memory and external storage, installed software, etc. The services, deployed on a host, determine the demand for these resources. If the cumulated demand for a resource of all the services deployed in a node exceeds the available level, then all the services will be degraded. Similarly, the logical connections between the services determine the demand on the communication resources at both end nodes of the connection. The network SLAs (Service Level Agreements) determine the limits placed on the cumulated communication demands in any single node or group of nodes. Thus, any change in the placement of services onto hosts affects both the time of processing requests by the services and the time of transmitting requests and responses. The problem of predicting this degradation is nontrivial, a simulation based approach is proposed in “Prediction of the performance of Web based systems” in this monograph.

3.2 Operational Configurations

System configuration is determined by the deployment of service components onto the system hosts. A configuration ensures system operability if the services are so deployed that the hosts are not overloaded and the demand for communication between them does not violate the SLA limits. The set of all possible configurations that satisfy these conditions is denoted by $\mathbf{\Psi}_{up}$. This set is referred to as the set of permissible configurations. It should be noted that some deployments will not be possible due to conflicting requirements of the services regarding the host resources, such as the versions of installed software. The corresponding configurations will also be excluded from the set of permissible ones.

Reconfiguration (change of system configuration) takes place when service deployment is changed. If we reconfigure the system to any configuration from the set $\mathbf{\Psi}_{up}$, then its operability will be preserved. Of course, this does not mean that the quality of the service will not be affected. The various permissible configurations may differ in the efficiency of generating the responses to client requests. This leads to the degraded operation after some reconfigurations. The set $\mathbf{\Psi}_{up}$ is then split to the disjoint sets of $\mathbf{\Psi}_l$ corresponding to the various levels of degradation $l \in [1..L]$.

The permissible and degraded-operation configurations can be found using standard combinatorial techniques and simulation. Due to the size of the problem, it is almost never feasible to compute the full sets, though.

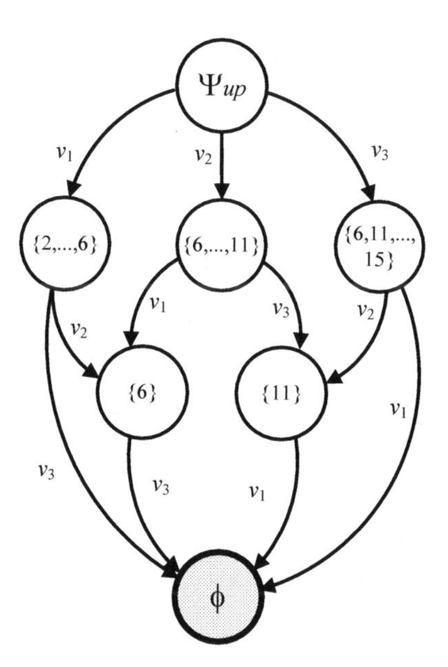

Fig. 3. An example of a simple reconfiguration graph (the numbers in the nodes correspond to the arbitrary numbering of permissible configurations in Table 1)

3.3 System Reconfiguration Graph

The reconfiguration graph [4] is built to define the possible changes in the configuration, that tolerate the various discussed faults. Set $\mathbf{\Psi}_{up}$ is at the root of the graph, since any admissible configuration ensures system being up, if there are no failures. The branches leaving the root correspond to the various faults affecting hosts or services. They point at subsets of $\mathbf{\Psi}_{up}$ obtained by eliminating the configurations which do not ensure system operation in presence of the specified faults, i.e. if a host is down as the effect of the fault occurrence, then all the configurations that assume deployment of a service to that host are eliminated. Let's denote the subset – obtained for each fault v_i – as $\mathbf{\Psi}|v_i$.

Further branches of the graph, corresponding to subsequent faults, are produced by eliminating configurations from $\mathbf{\Psi}|v_i$. These are denoted as $\mathbf{\Psi}|v_1v_2$. The procedure is continued until the elimination produce empty sets ϕ that correspond to combinations of failures that cannot be tolerated by any reconfiguration. This approach to the reconfiguration graph construction ensures that all the possible configurations are taken into account. An example of such a reconfiguration graph is presented in Fig. 3.

It should be noted that the reconfiguration graph illustrates all the possible changes in the service deployment that will preserve the system operability.

3.4 Testbed Analysis

Let's consider a fairly simple system to illustrate the proposed approach to dependability analysis. The system consists of 3 hosts: Server A, Server B and Server C. There are also 3 service components: Controller, View and Model. Table 1 lists the permissible deployments of the services onto hosts. Configuration 1 is used when the system is fully operational.

Table 1. Permissible configurations

Id.	Controller	View	Model
1	Server A	Server B	Server C
2	Server B	Server B	Server C
3	Server B	Server C	Server B
4	Server C	Server B	Server C
5	Server C	Server C	Server B
6	Server C	Server C	Server C
7	Server A	Server A	Server C
8	Server A	Server C	Server C
9	Server C	Server A	Server C
10	Server C	Server C	Server A
11	Server A	Server A	Server A
12	Server A	Server B	Server A
13	Server A	Server A	Server B
14	Server B	Server A	Server B
15	Server B	Server B	Server A

When a fault occurs, one of the hosts becomes unavailable. Then, some of the configurations cannot provide service anymore. The list of permissible configurations after any combination of faults is enumerated in Table 2. This is the basis for constructing the reconfiguration graph in Fig. 3.

Table 2. Admissible configurations for the given set of faults

Fault Set	Configurations
v_1	2,3,4,5,6
v_2	6,7,8,9,10,11
v_3	6,11,12,13,14,15
v_1, v_2	6
v_3, v_2	11

The state-transition graph can be directly derived from the reconfiguration graph. All the up-states of the S-T graph are represented as nodes of the reconfiguration graph. The fail-states are combined to a single state. The Markov chain is then derived by annotating the transitions with intensities of the faults occurrence λ_i. Further transitions have to be added, corresponding to the renewal/repair process being implemented. Renewal can be realized independently for each fault. This is represented by the transitions directed opposite to every fault occurrence transition, characterized by the corresponding renewal intensity μ_i. An extra transition is introduced from the fail-state to the fully operational system – this represents a global renewal policy with intensity M . The resulting Marcov chain is represented in Fig. 4.

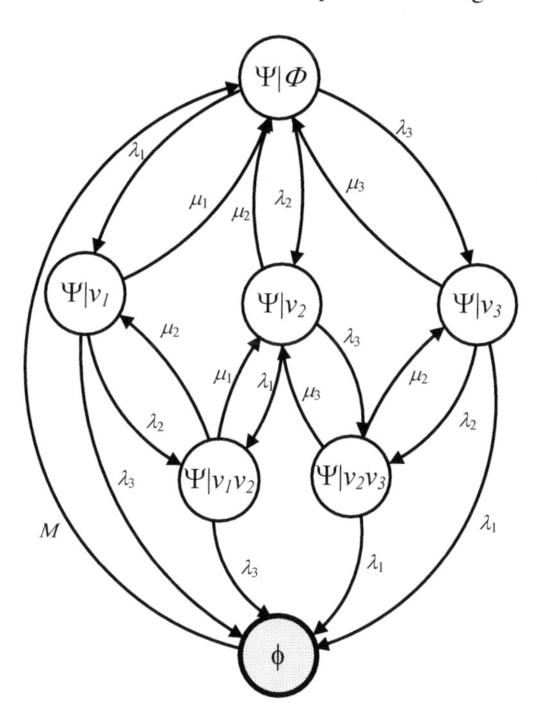

Fig. 4. An example of a the S-T model for reconfiguration graph from Fig. 3

The steady state probabilities of the state transition model given in Fig. 4 can be determined by solving the following set of equations:

$$\Lambda^T \circ \begin{bmatrix} P\{\Psi \mid \Phi\} \\ P\{\Psi \mid v_1\} \\ P\{\Psi \mid v_2\} \\ P\{\Psi \mid v_3\} \\ P\{\Psi \mid v_1 v_2\} \\ P\{\Psi \mid v_2 v_3\} \\ P\{\phi\} \end{bmatrix} = \begin{bmatrix} 0 \\ 0 \\ 0 \\ 0 \\ 0 \\ 0 \end{bmatrix}, \tag{5}$$

$$\begin{aligned} &P\{\Psi \mid \Phi\} + P\{\Psi \mid v_1\} + P\{\Psi \mid v_2\} + P\{\Psi \mid v_3\} + \\ &P\{\Psi \mid v_1 v_2\} + P\{\Psi \mid v_2 v_3\} + P\{\phi\} = 1 \end{aligned} \tag{6}$$

where:

$$\Lambda^T = \begin{bmatrix} -\begin{pmatrix} \lambda_1+\lambda_2 \\ +\lambda_3 \end{pmatrix} & \mu_1 & \mu_2 & \mu_3 & 0 & 0 & M \\ \lambda_1 & -\begin{pmatrix} \lambda_2+\lambda_3 \\ +\mu_1 \end{pmatrix} & 0 & 0 & \mu_2 & 0 & 0 \\ \lambda_2 & 0 & -\begin{pmatrix} \lambda_1+\lambda_3 \\ +\mu_2 \end{pmatrix} & 0 & \mu_1 & 0 & 0 \\ \lambda_3 & 0 & 0 & -\begin{pmatrix} \lambda_1+\lambda_2 \\ +\mu_3 \end{pmatrix} & 0 & \mu_2 & 0 \\ 0 & \lambda_2 & \lambda_1 & 0 & -\begin{pmatrix} \lambda_3+\mu_1 \\ +\mu_2 \end{pmatrix} & \mu_3 & 0 \\ 0 & 0 & \lambda_3 & \lambda_2 & 0 & -\begin{pmatrix} \lambda_1+\mu_2 \\ +\mu_3 \end{pmatrix} & 0 \\ 0 & \lambda_3 & 0 & \lambda_1 & \lambda_3 & \lambda_1 & -M \end{bmatrix}. \tag{7}$$

The equations are solved by linear programming and the steady state probabilities are found. This is the basis for determining the system availability from equation (3).

To simplify the algebraic solution of equations (3, 5-7), the intensities of the faults occurrence are assumed equal to λ for all the groups. Also, the local renewal intensities are similar, denoted as μ, and global renewal policy equal to $\mu/2$. For such simplifications the availability is then given as:

$$A = \frac{\mu\left(4\mu^4 + 58\lambda^3\mu + 59\lambda^2\mu^2 + 25\lambda\mu^3 + 18\lambda^4\right)}{24\lambda^4 + 90\lambda^4\mu + 118\lambda^3\mu^2 + 75\lambda^2\mu^3 + 25\lambda\mu^4 + 4\mu^5}. \tag{8}$$

Fig. 5.A presents the results of availability analysis computed from (8) for various values of the local renewal and faults occurrence. The impact of fault rate and renewal time is as expected for any repairable system. It is much more interesting to compare these results with availability of a system where we choose not to relocate services. Let us define reconfiguration availability improvement factor as a ratio of availability change over availability of the system without reconfiguration ($\overline{A}$):

$$r = \frac{A - \overline{A}}{\overline{A}} \cdot 100\% \tag{9}$$

The results of reconfiguration improvement factor computed from (9) for the local renewal and faults occurrence values the same as in the previous analysis are presented in Fig 5.B. In the analyzed area of the mean renewal and failure time the use of service reconfiguration improves the web system availability up to 20%. The improvement is meaningless for very fast renewal times.

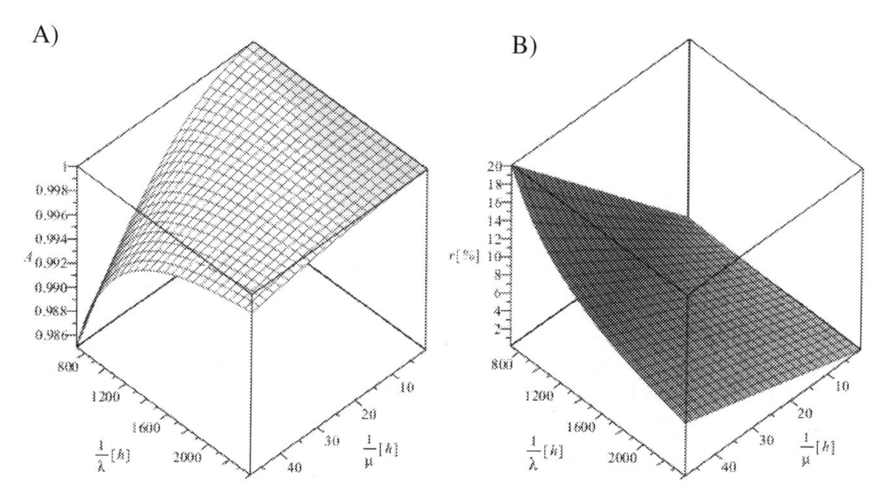

Fig. 5. System availability of a web system employing reconfiguration and B. Reconfiguration availability improvement factor when a web system uses reconfiguration to improve dependability.

4 Optimization of the Reconfiguration Strategy

The reconfiguration graph describes all the possible target configurations that can assure continuity of the services after a sequence of faults occurrence. This is done by following the graph transitions from its root, corresponding to the faults sequence. The set of configurations, associated with the vertex thus reached, represents all the target configurations that tolerate this faults sequence. If the set is empty, then the faults cannot be tolerated and the system fails.

Reconfiguration strategy is constructed by choosing just one configuration from the set in each node of the reconfiguration graph. If any of the graph nodes (except for the root one) contain multiple alternate configurations, then there are different reconfiguration strategies that can be constructed in this way.

Any one of the so obtained strategies is equivalent from the point of view of service availability, since the equation (3) does not depend on the choice of configurations in the reconfiguration graph nodes. The proposed measures of system resilience (i.e. single points of failure *SPF* and multiple points of failure *MPF*) are also invariant with the choice of reconfiguration strategy, since they can be directly derived by analyzing the reconfiguration graph.

Thus, choosing the optimal strategy can be either based on the proposed quality of service measure (4) or may require additional criteria.

4.1 Reconfiguration Strategy Optimizing the Quality of Service

We consider the situation, where the optimal strategy should ensure the highest overall quality, i.e. the quality given by equation (4) has to be maximized over the set of all permissible reconfiguration strategies. The choice of optimal strategy does not change the reconfiguration graph and, what follows, the S-T graph. This means that the state probabilities in (4) are invariant, i.e.

$$\max Q = \sum_{s \in \mathbf{S}_{up}} P_s \cdot \max_{k \in \Psi_s} \{q_k\} \tag{10}$$

This means that the problem of finding the optimal strategy can be reduced by determining the configuration with the highest quality coefficient in each node of the reconfiguration graph.

The problem can be further simplified considering that some probabilities P_s are much smaller than the others. Usually, this corresponds to nodes of reconfiguration graph reached after multiple independent errors (though not necessarily). The choice of configuration in these nodes does not impact the overall quality significantly. In these nodes it is sufficient to determine any permissible configuration, not necessarily the best one.

This approach to determining the reconfiguration strategy has to be based on some measure of the quality of services. This is always questionable, as there are multiple metrics to be considered. Strategies that maximize one, usually do not perform so well in the others. Often, the only reasonable approach is to classify the system performance into a few categories – levels of performance degradation – and base optimization on achieving the best average level of degradation (an approach based on arbitrary classification based on expert knowledge).

A more formal approach may be based on the analysis of response time characteristics of business services for the various configurations. See Fig. 6 for an example of the response time characteristic. Obviously, it is a function of the demand for service and not just a single number that could be used in equation (4). It has a distinguishing attribute: the maximum properly handled request rate. This can be used as the measure of service quality.

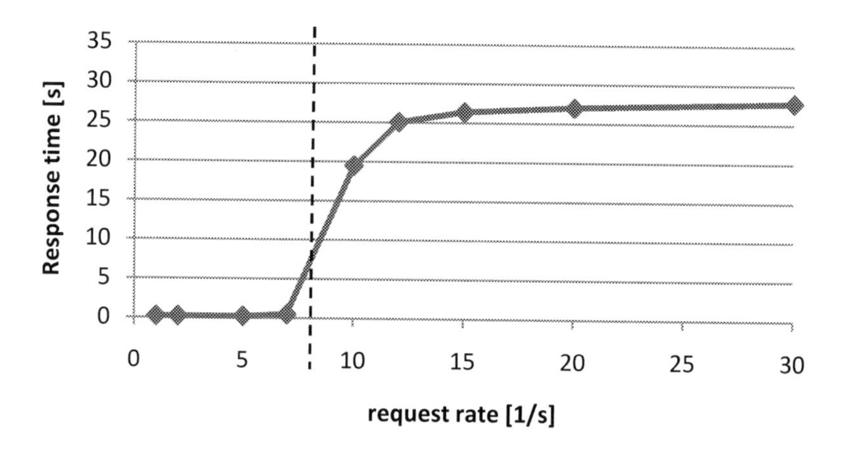

Fig. 6. Response time characteristic of a web based system configuration

4.2 Reconfiguration Strategy Minimizing Service Relocation

Optimizing the reconfiguration strategy on the basis of quality of service often does not yield significant improvement in the overall quality. This is the case when the faults are rare (their probabilities are small). In such situations, it may be more important to address some other aspects connected with reconfiguration. Obviously, the number of service components that have to be relocated may be an important issue.

Let's consider an example of the reconfiguration strategy shown in Fig. 7. By comparing the configurations in the neighboring nodes of the graph (connected by a transition), it is possible to determine the number of service components n_i that have to be relocated during reconfiguration. The number is then used to describe the corresponding transition. By repeating it for every transition in the graph, the marking shown in Fig. 6 is obtained. It illustrates the numbers of relocated service components in every situation envisioned by the strategy. Of course, the probability of reconfiguration has also to be considered during optimization. Thus, the component numbers at each transition are weighed by the probabilities of states from which they initiate. The sum of these attributes, taken over all the graph transitions, is further called the average relocation factor. It can be used to optimize the service strategies to account for difficulty of reconfiguration.

It should be noted that changing the configuration in a single node, affects the weights of all the transitions reaching or leaving that node. In effect, the optimization process cannot be limited to single nodes, but must consider the whole strategy. Any of the well-known nonlinear optimization techniques can be applied.

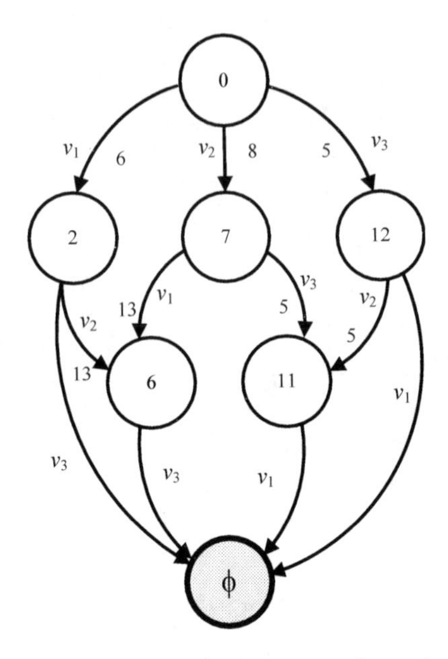

Fig. 7. An example of service relocation assessment for a relocation strategy

5 Conclusions

We demonstrated how a properly developed reconfiguration strategy can improve the various aspects of dependability (preserving continuity of service, removing or reducing the number of single points of failure, improving over-all availability, optimizing the quality of service). The proposed technique involves identification of the potential faults, construction of the reconfiguration graph, assessment of the steady state probabilities of web system dependability states, choosing the optimal strategy from among the admissible ones (identified by the reconfiguration graph).

The proposed approach is illustrated in a very simple case study: a web system consisting of just 3 services deployed to 3 hosts. The resulting reconfiguration graph is still quite complex. In real world systems, the graph can be huge and the corresponding strategies very complex. In these cases, the formulated optimization problems require application of some formal algorithms. These are addressed elsewhere in this monograph.

Acknowledgements. The presented work was funded by the Polish National Science Centre under grant no. N N516 475940.

References

1. Avizienis, A., Laprie, J., Randell, B.: Fundamental Concepts of Dependability. In: Proc. 3rd IEEE Information Survivability Workshop, Boston, Massachusetts, pp. 7–12 (2000)

2. Barlow, R.E.: Engineering Reliability. ASA-SIAM Series on Statistics and Applied Probability (1998)
3. Caban, D., Walkowiak, T.: Dependability oriented reconfiguration of SOA systems. In: Grzech, A. (ed.) Information Systems Architecture and Technology: Networks and Networks' Services, pp. 15–25. Oficyna Wydawnicza Politechniki Wrocławskiej, Wroclaw (2010)
4. Caban, D., Zamojski, W.: Dependability analysis of information systems with hierarchical reconfiguration of services. In: Second International Conference on Emerging Security Information, Systems and Technologies, SECURWARE, pp. 350–355. IEEE Press (2008)
5. Eusgeld, I., Freiling, F.C., Reussner, R. (eds.): Dependability Metrics. LNCS, vol. 4909. Springer, Heidelberg (2008)
6. Krekora, P., Caban, D.: Dependability analysis of reconfigurable information systems. In: 2nd International Conference on Dependability of Computer Systems, DepCoS-RELCOMEX, pp. 177–184. IEEE Press (2007)
7. Musa, J.D.: Software Reliability Engineering. More Reliable Software, Faster Development and Testing. McGraw-Hill (1999)
8. Tartanoglu, F., Issarny, V., Romanovsky, A., Levy, N.: Dependability in the Web Services Architecture. In: de Lemos, R., Gacek, C., Romanovsky, A. (eds.) Architecting Dependable Systems. LNCS, vol. 2677, pp. 90–109. Springer, Heidelberg (2003)
9. Pérez, P., Bruyère, B.: DESEREC: Dependability and Security by Enhanced Reconfigurability. European CIIP Newsletter 3(1) (2007)
10. Walkowiak, T.: Information systems performance analysis using task-level simulator. In: DepCoS – RELCOMEX, pp. 218–225. IEEE Computer Society Press (2009)
11. Walkowiak, T., Michalska, K.: Functional based reliability analysis of Web based information systems. In: Zamojski, W., Kacprzyk, J., Mazurkiewicz, J., Sugier, J., Walkowiak, T. (eds.) Dependable Computer Systems. AISC, vol. 97, pp. 257–269. Springer, Heidelberg (2011)

Functional-Reliability Model of a Services System with Path Reconfiguration Ability

Wojciech Zamojski and Jarosław Sugier

Wrocław University of Technology, Wrocław, Poland
Institute of Computer Engineering, Control and Robotics
{wojciech.zamojski,jaroslaw.sugier}@pwr.wroc.pl

Abstract. In this work we analyse operation of a services network which is built from processing nodes serving dedicated services and communication links transmitting required information resources. For task executions specific subsets of network resources – so called communication paths – are allocated. Execution of a task in communication paths is disrupted by occurring faults which are eliminated by renewal procedures leading to delays in task completion time or even to task cancellation. One of the method for overcoming the negative effects of the faults is to apply a path reconfiguration mechanism i.e. a redirection of the communication traffic which bypasses the damaged link. We propose a functional-reliability dependability model of the services network which takes into account possible path reconfigurations. We also define a network dependability parameter which evaluates network efficiency by finding the degree of task losses among all the jobs being executed in the network.

Keywords: dependability, services network, dependability model of a services network, communication path, reconfiguration, path reconfiguration.

1 Introduction

In this paper we investigate networks of services which are constructed from processing nodes (providing dedicated services) and communication links (ensuring proper transmission of data related to both tasks being handled and operation the of network itself). Realized tasks engage particular network resources and for any given task there can be several different hardware, software or functional configurations which accomplish it with different efficiency (throughput) and along different time schedules [11, 13]. The last observation is particularly significant when secure and timely completion of the tasks (services) is required in a network built from unreliable (fallible) components (nodes and communication links) and operating in not always friendly environment. Such an environment can often be a source of threats and attacks of random or deterministic (purposeful) nature [1].

A good example of a services network can be a computer network with an architecture particularly well suited for execution of user's (client's) tasks based on dynamically allocated functionalities – so called Service-Oriented Architecture, SOA. In

W. Zamojski and J. Sugier (eds.), *Dependability Problems of Complex Information Systems*,
Advances in Intelligent Systems and Computing 307, DOI: 10.1007/978-3-319-08964-5_10

these kinds of systems it is usually possible to substitute particular functionalities with other equivalent ones implemented with different hardware / software configurations or completed in different time [15, 13, 16].

The resource allocation process is inherently dynamic and is determined by the events of tasks being submitted, resources being released, occurring faults, activations of system recovery or failover procedures, etc. The issues analysed in this work belong to the topics of advanced contemporary information technologies like the above mentioned SOA (Service Oriented Architecture), SOM (Service Oriented Management) or SCA (Service Component Architecture) which deal with problems of service / resource allocation and selection of exploitation strategy which would be optimal for user demands like it is, for example, in complex e-business systems [15, 16, 7].

2 Services Networks

A services network considered in this paper is interpreted as a server – client system which realizes user (client) tasks in a collection (farm) of servers and hosts.

To some degree there is an analogy between the model of communication paths which is introduced below and TCP/IP computer networks, mobile wireless networks or cloud computing systems.

2.1 Communication Structure of the Network

To accomplish a requested task a communication path is established which comprises necessary network resources. The path consists of the transmitter node, a set of intermediate nodes, the receiver node and required communication links. It is assumed that functionalities of particular nodes are activated dynamically. In many practical cases functionality of any intermediate node is limited to setting up a communication path understood as pointing to the nearest node which is active and operational i.e. it can confirm correct reception of transmitted messages [7].

In order to precisely investigate the functional-dependability network model the following concepts are hereby introduced [12, 13]:

- The transmitter and the receiver nodes are considered to be *neighbour nodes* if it is possible to correctly transmit information (messages, packets) between them.
- The receiver node is the *n-th order neighbour* of the transmitter node if active and correct operation of $n - 1$ intermediate nodes and n links is required for proper transmission between them.
- The *distance* between any two nodes of the network is defined as the overall time required for transmission between them.

The *transmission time* covers overall time of information transfer over the transmission medium, i.e. the time needed for establishing (setting up) the communication path between the sender and the receiver and the actual time of the correct transmission itself. The higher the order of node neighbourhood, the longer the distance between them, as understood in the above definition.

2.2 The Tasks

The process of completing a task includes:

- Task feasibility analysis, i.e. testing whether the network has required services at its disposal. As a result of this phase the task is accepted for execution or rejected (refused).
- Preparation of task execution choreography, i.e. organization of a chain of services which also includes allocation of necessary resources from informational and technical infrastructures. Allocation of services and resources can be static or dynamic – the latter one is implemented, among others, with balancing the load placed upon the nodes (servers) and communication links.
- The proper completion of the task, i.e. executing in the request – respond mode consecutive elements of the service chain where each particular element of the chain triggers sending data packets over communication links to service providers. In case of communication disruptions the network tries to repeat particular transmissions or even attempts to reconfigure the communication paths.

It is assumed that the task completion time is the sum of individual service times and transmission times which occur in the communication paths;

$$\tau_T^{(i)} = \sum_{services} \left(\tau_{serv}^{(i)} + \tau_{comm}^{(i)} \right) \tag{1}$$

where

$\tau_{serv}^{(i)}$ - execution time for the i-th service,

$\tau_{comm}^{(i)}$ - communication procedures time related to execution of the i-th service.

In a real environment execution times for both services and communication procedures depend on numerous factors, including faults and disruptions, and are random. In practice these times are either estimated with expected values (possibly medians) or intervals of their variation are evaluated.

3 Dependability

Dependability is a property of a system (a network, an object) which integrates such attributes as perfomability, reliability, readiness, security, survivability and maintenance – all related to correct and in-time execution of the tasks [1, 14].

Dependability analysis takes into account, among others:

- threats, faults and errors which occur in technical structures and management systems;
- functional and performance characteristics;
- actions which reduce consequences of occurred or foreseen threats, faults or errors.

3.1 Errors and Faults

There are many causes of faults in information systems and among them hardware failures are now becoming more and more infrequent and insignificant. Today the main source of faults in computer systems are errors which are brought into system operation by software and by people (administrators, operators, users). The nature of these faults is also changing; "classic permanent" failures calling for a repair (technical renewal) are becoming rare while transient faults (misrepresentations or errors) are becoming frequent. Although malfunctions (e.g. transmission faults) or errors do not demand repairs, they can cause significant time losses as they require reconstruction of the interrupted processing procedures, i.e. the informational renewal [11, 9].

Fault of a communication path is defined as a random event of breaking a communication link between a pair of neighbour nodes. A direct cause for such a fault can be an event from the following categories:

- *F1* – a set of physical failures of the links,
- *F2* – a set of node failures (terminal and/or intermediate ones),
- *F3* – a set of events related to overloading the communication paths with excessive number of transmitted messages which leads to exceeding the limits of service completion time (queue problems) and rejecting the subsequent packets (messages).

3.2 Renewals

Renewal of a communication path is understood as a restoration of its functional-reliability parameters by;

- *r1* – removing the physical failure of the link (repair) – $r1_H$ and re-establishing the communication path (restoring the original functional properties of the originally established path) – $r1_S$,
- *r2* – removing the physical failure of a node (repair) and re-establishing the communication path (restoring the functional properties)
- *r3* – reconfiguration of the communication path understood as redefinition of node neighbourhood and redirection of the traffic to the "nearest" node.

3.3 Strategies for Restoration of Communication Paths

The following strategies for restoration of communication paths are defined.

1. Repair or reconfiguration of a physically damaged link
As a rule, a repair is a prolonged operation and its completion time significantly surpasses average task completion time in the network ($\tau_{r1H} >> \tau_T$). In networks with built-in path reconfiguration mechanisms a by-pass link is created for the time of restoration. Cost of establishing a by-pass link (of usually smaller throughput) is

evaluated from the time required for its arrangement. In this case the restoration time meets the conditions

$$if\ \left(f^{(j)} \in F1\ \ and\ \exists\, reN^{(j)}\right)\ then\ \tau_{reN}^{(j)} \leq \tau_{REN}^{(i)} \leq \tau_{r1H}^{(j)} + \tau_{r1S}^{(j)} \cong \tau_{r1H}^{(j)} \tag{2}$$

where

$reN^{(j)}$ - path reconfiguration triggered by the j-th fault,

$\tau_{REN}^{(j)}$ - renewal time,

$\tau_{reN}^{(j)}$ - time of path restoration,

$\tau_{r1H}^{(j)}$ - time of technical renewal (repair) of the link,

$\tau_{r1S}^{(j)}$ - time of functional restoration of the path (informational renewal).

2. Repair or reconfiguration of a physically damaged node
Like it was in the case of a link damage, a node damage is eliminated by a repair ($r2H$) or a reconfiguration which takes $\tau_{reN}^{(j)}$ time, i.e.

$$if\ \left(f^{(j)} \in F2\ \ and\ \exists\, reN^{(j)}\right)\ then\ \tau_{reN}^{(j)} \leq \tau_{REN}^{(j)} \leq \tau_{r2H}^{(j)} + \tau_{r2S}^{(j)} \cong \tau_{r2H}^{(j)} \tag{3}$$

where

$\tau_{r2H}^{(j)}$ - node renewal (repair) time,

$\tau_{r2S}^{(j)}$ - time of functional recovery of the path.

3. Elimination or reduction of paths overloading
Overloading of a communication path comes as a consequence of throughput of the nodes or links being too low compared to the number of messages sent and it can lead to accumulation of the messages in node buffers. Pending packets can be sent farther with different, "substitute", links (provided that the network has appropriate reconfiguration abilities) or they can be eliminated from transmission if the packet waiting time exceeds permissible waiting time (task queuing). Elimination of the overloads – the bottlenecks – is a procedure which includes setting up a substitute (of worse efficiency) path and accomplishing the transmission itself. In many cases completion time for these operations can be comparable to the average task execution time ($\tau_{r3S}^{(j)} \approx \tau_T$) whereas the cost analysis should consider overall reduction in network performance resulting from redirection of the flow to the substitute paths (which are already loaded with other assigned tasks) [2, 10].

3.4 Measures of Network Dependability

As a measure of network dependability we can take an effectiveness factor defined as a ratio of actually executed tasks $n_{real\,T}$ to the potential performability [11, 12,1] of the network n_T which represents a number of tasks completed in conditions of its ideal reliability and maximal productivity

$$\eta_T = \frac{n_{realT}}{n_T} \approx \frac{\tau_{nomT}}{\tau_{realT}}\left(1 - \frac{T_{lost}}{T}\right) \tag{4}$$

where

T – length of analysed time horizon of network operation,

$\tau_{nom\,T}$ – nominal time of task completion,

$\tau_{real\,T}$ – actual time of task completion,

T_{lost} – time spent (lost) on recovery after failures and their repercussions.

The number of tasks accomplished in considered time horizon [0, *T*] and completion times of particular tasks depend on functional and performance properties of the network, faults disrupting its operation and organization of processes for renewal of its hardware and informational resources.

4 Impact of Communication Links Reconfigurability on Network Dependability

4.1 Introduction

In the forthcoming analysis we will consider a network with *N* services implemented in dedicated nodes. For simplicity it is assumed that the services assigned to the *i*-th specific task are numbered from 1 to $N^{(i)}$ ($N^{(i)} \le N$) and are carried out sequentially according to this numbering. The task is initiated in the first node and finished in the node $N^{(i)}$. Implementation of the task require transmissions which incur packet transfer times and some time used for setup of the communication path.

In many cases, it is assumed that during the implementation of the *i*-th task there will be not more than one physical damage of the connection between two neighbouring (*j*, *j+1*) nodes. From the commonly accepted principles of the reliability theory probability of such an event is approximately equal to $\Pr\{f_{j,j+1}, t \le \tau_T^{(i)}\} \cong \lambda^1_{j,j+1}\tau_T^{(i)}$. Link repair time usually by far exceeds the average execution time of network tasks. If ability to reconfigure communication paths is built-in into the network then for the time of link renewal a substitute connection is established which bypasses the damaged one under repair (Figure 1). The cost of establishing a substitute path/link, usually with lower bandwidth, is evaluated from amount of time spent on this operation.

4.2 Reconfiguration Costs

Let's consider a communication path from the transmitting node S to the receiver R (Figure 1). Detection of a damaged link between nodes i and j starts the path reconfiguration procedure for this link. Probably the most profitable network reconfiguration will be the one using the first order neighbour node, i.e. $i \overset{b}{\rightarrow} j \Rightarrow (reN = 1): i \rightarrow (i+1) \rightarrow j$ (cf. Figure 1). Other possible reconfigurations, for example $(reN = 2): i \rightarrow (i+1) \rightarrow (i+2) \rightarrow j$, involve a larger number of intermediate links and nodes, and thus will be more expensive (longer delivery time).

Estimation of the value of the network tasks loss factor (4) in the interval of time T can be carried out on the basis of the expected values and / or the lower and upper limits of variation of $T_{lost}^{(i)}$ as well as $\tau_{realT}^{(i)}$.

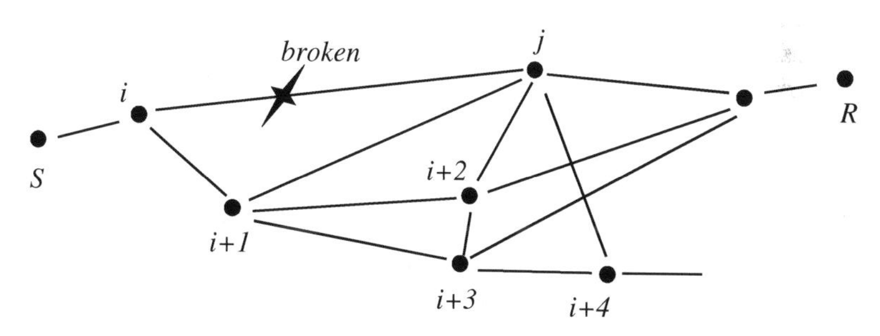

Fig. 1. Fragment of a network with a broken link $i \rightarrow j$

4.2.1 Time Lost T_{lost}

It is assumed that the network executes a stream of i-th tasks with a total duration time $T^{(i)} = \sum_i \tau_T^{(i)}$ and ($T^{(i)} >> MTBF^{(i)} >> \tau_{realT}^{(i)} \geq \tau_{nom\,T}^{(i)}$), and during this time $F^{(i)}$ faults of communication links/paths occur that are recovered (recovery time $\tau_{REN}^{(i)}$). In the absence of path reconfiguration, accepting an assumption of failures independence, the time lost in use of the communication path corresponds to a total time of repairing the damaged resource:

$$T_{lostT}^{(i)} \approx \sum_{f^{(l)}} \Pr\left\{f^{(l)}; \left(T^{(i)} - T_{lostT}^{(i)}\right)\right\} * \Pr\left\{\overline{reN}\ ; f^{(l)}\right\} * T * \tau_{REN}^{(i)} \tag{5}$$

where

$\Pr\left\{f^{(l)}; \left(T^{(i)} - T_{lostT}^{(i)}\right)\right\}$ - probability of the l-th failure occurring in actual time of task execution,

$\Pr\left\{\overline{reN}\ ; f^{(l)}\right\}$ - probability of the l-th fault being unresolvable through path reconfiguration.

Calculation of the lost time on the basis of equation (5) is "difficult in computational practice" so it is proposed to simplify the problem and to use estimates of the expected values and / or limits of variation ranges.

We will use notation $E_T[...]$ to represent the expected value operator applied to the interval [0, *T*]. For example, if it is assumed that distribution of the lifetime of a path with *N* resources (reliability serial configuration) is described by exponential distribution with failure rate λ_{comm} than the following estimation of lost time is possible

$$0 \leq \lambda_{comm}\, T\, \tau_{REN} \leq E_T\left[T_{lost}\right] \leq N\lambda_{comm}\, T\, \tau_{REN} \tag{6}$$

It is easy to see that both (the upper and the lower) estimated limits of the time lost are too pessimistic because they do not take into account the basic observation that during the renewal process some functionalities of the system may be "switched off" and therefore they are not subject to damage. Thus, $(T - T_{lost})$ instead of T should be taken into account. The bottom (left-hand) estimation of the time lost is closer to reality because the actual MTBF times of hardware (servers and communication links) are large enough for the probability of more than one being damaged during such a long time period (one or two years of use) becoming negligible. In our opinion given estimation of the upper time lost variation limit is too pessimistic (the worst case), especially because it is made against additional assumptions about sequential delivery of services.

4.2.2 Duration of the Tasks $\tau_{realT}^{(i)}$

The nominal (without disruptions brought by network failures) execution time of a task using the functionalities available in the network is defined by the relation (1). Assuming that execution times of services and communication times shall be the same we get

$$\tau_T^{(i)} = N_{serv}^{(i)} \tau_{serv} + N_{comm}^{(i)} \tau_{comm}^{(i)} \tag{7}$$

A damage of the communication link starts (concurrently to the repair/renewal process) a reconfiguration process of the communication path which includes 1) determining necèssity of a new path - an additional load on the processor operating the service, 2) "extension" of the transmission path through additional intermediate links and nodes. If the issue of network bandwidth changes is ignored, we obtain

$$\tau_{realT}^{(i)} \approx \tau_T^{(i)} + \Pr\left\{reN^{(i)}\right\}\left(\tau_{serv} + 2\tau_{comm}^{(i)} + \tau_{reN^{(i)}}\right) \tag{8}$$

where

$\Pr\left\{reN^{(i)}\right\}$ - probability of reconfiguration,

$\tau_{reN^{(i)}}$ - reconfiguration time.

Path reconfigurations are accomplished based on available and efficient resources (the connecting links and the nodes). The lower estimation of the probability of successful reconfiguration corresponds to the condition when a service is transferred to one of unused resources treated as cold reserve, and the upper estimation – to the situation when all the remaining nodes and links get involved in this operation. Our further discussion we will be limited to the networks with reconfiguration groups of K paths in the form link – server – link which in the reliability theory is modelled as a system with $N^{(i)}$ serially connected elements with a moving loaded reserve of cardinality K. [12, 5].

The above estimations of the task duration time (7), the probability of a successful reconfiguration of the path (8) and the lost time (5) are modified depending on a particular strategy of reconfiguration implemented in the network.

4.2.3 Effectiveness Factor of a Network with Reconfiguration of Communication Paths

In a case of a network with reconfigured communications paths the reconfiguration effectiveness factor (4) is modified to the form of

$$\eta_T^{(i)}(reN^{(i)}) = \frac{n_{realT}^{(i)}(reN^{(i)})}{n_{nom\,T}^{(i)}} \approx \frac{\tau_{nom\,T}^{(i)}}{\tau_{realT}^{(i)}(reN)}\left(1 - \frac{T_{lost}^{(i)}(reN^{(i)})}{T}\right) \tag{9}$$

where

$N^{(i)}$ – resources involving in realization of the i-th task,

$reN^{(i)}$ – used reconfiguration strategy.

4.3 Reconfiguration According to the Principle of the First-Order Neighbourhood

4.3.1 The *reN*=1 Reconfiguration

It is assumed that the network is built of N resources $N = N_S + N_C$ where N_S - a number of servers (nodes) of the network and N_C - a number of communication links. Realization of the i-th task is based on $N^{(i)} = N_S^{(i)} + N_C^{(i)}$ resources ($N^{(i)} \leq N$) which create a serial reliability configuration. All the servers and all the communication links have the same functional and reliability properties but the number of used processors is larger by one than the number of connecting links. The first-order neighbourhood reconfiguration strategy ($reN^{(i)} = 1$) is used which is based on free resources i.e. the ones not involved in realization of the task process.

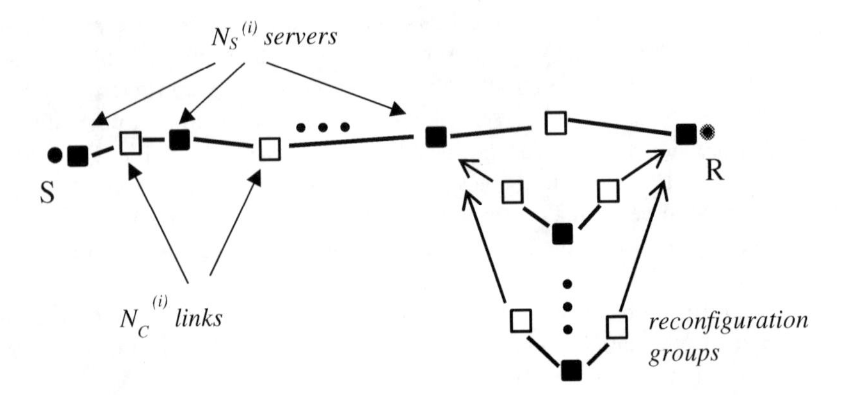

Fig. 2. Reliability schema of the *i*-th task functional configurations with a reconfiguration group implementing strategy $reN^{(i)} = 1$

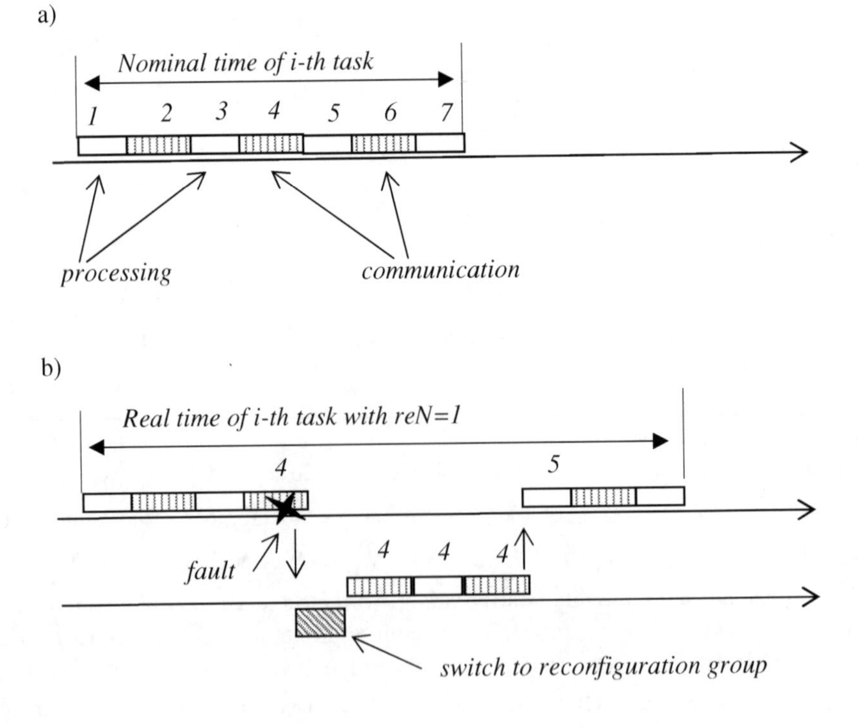

Fig. 3. Functional schema of the *i*-th task realization; a) no faults, b) a fault and a reconfiguration with strategy $reN^{(i)} = 1$

If there are no faults in functional configuration during the period of i-th task execution $F\left(N^{(i)}=N_S^{(i)}\cup N_C^{(i)}\right)=0$ then the task completion time is equal to the nominal duration time of a task (1) and it is estimated as $\tau_{nomT}^{(i)}\approx N_S^{(i)}\tau_{serv}+N_C^{(i)}\tau_{comm}$. On the other hand, if there is one functional configuration fault $F\left(N^{(i)}=N_S^{(i)}\cup N_C^{(i)}\right)=1$, then the actual task completion time in the network with reconfiguration according to the principle of first-order neighbourhood ($reN^{(i)}=1$) is associated with involvement of one additional server node and two communication links and the time of task realization may be generally estimated as a reconfiguration procedure (see Figure 3);

$$\begin{aligned}\tau_{realT}^{(i)}\left(reN^{(i)}=1\right)\approx \Pr\left\{F\left(N^{(i)}\right)=0\right\}\left(N_S^{(i)}\tau_{serv}+N_C^{(i)}\tau_{comm}\right)+\\+\Pr\left\{reN^{(i)}=1\right\}\left(\tau_{nomT}^{(i)}+\tau_{serv}+2\tau_{comm}+\tau_{reN^{(i)}}\right)\end{aligned}\tag{10}$$

where

$\Pr\left\{F\left(N^{(i)}\right)=0\right\}$ - probability of correct execution of the i-th task with resources of the basic functional configuration (without consuming any resources of the reconfiguration group),

$\Pr\left\{reN^{(i)}=1\right\}$ - probability of proper completion of the task with successful application of the resources from the reconfiguration group.

It is assumed that, for increasing the probability of successful completion of the i-th task, a reconfiguration group of cardinality K has been created which operates according to the principle of first-order neighbourhood ($reN^{(i)}=1$). Cardinality of a such reconfiguration group significantly affects the actual task execution time – see Table 1.

4.3.2 Network Cases

Let's denote probability of correct operation of the server as r_S and of the communication link as r_C. It is also assumed that detection of a damaged communication path starts a network renewal process which comprises both the reconfiguration process and the repair processes for the failed resources.

<u>*Case 1*</u>

Probability of operation of the network in this case equals probability of successful completion of the task in functional configuration $N^{(i)}$ with serial reliability structure, i.e.

$$F\left(N^{(i)}\right)=0\Rightarrow P_{Case1}^{(i)}=r_S^{N_S^{(i)}}\,r_C^{N_C^{(i)}}\left(r_S\,r_C^2\right)^{K^{(i)}}\tag{11}$$

and the time network remains in this case corresponds to the nominal task execution time

$$\tau_{Case1} \approx \tau_{nom\,T}^{(i)} = N_S^{(i)}\tau_{serv} + N_C^{(i)}\tau_{comm} \tag{12}$$

Table 1. Considered cases of network reconfiguration

Case	*Events*	*Description*
1	$F\left(N^{(i)}\right)=0 \cap F(REC)=0$	Correct execution of the task with resources of the functional configuration $N^{(i)}$ and with fully operational reconfiguration group
2	$F(N^{(i)}) = 1 \cap F(REC) = 0$	Failure of one of the resources from the functional configuration $N^{(i)}$ and successful reconfiguration of the network with presumed correct operation of all resources from the reconfiguration group; reconfiguration time $\left(\tau_{serv} + 2\tau_{comm} + \tau_{reN^{(i)}}\right)$
3	$F(N^{(i)}) = 1 \cap F(REC) = k;$ $1 \le k < K$	Failure of one of the resources from the functional configuration $N^{(i)}$ and of k ($1 \le k < K$) paths in the reconfiguration group; reconfiguration time is estimated as $\tau_{serv} + (2k-1)\tau_{comm} + k\,\tau_{reN^{(i)}}$
4	$F(N^{(i)}) = 1 \cap F(REC) = K$	Failure of one of the resources from the functional configuration $N^{(i)}$ and of all K paths in the reconfiguration group, i.e. the i-th task cannot be completed and network renewal (which takes $\tau_{REN}^{(i)}$ time) is necessary

<u>*Case 2*</u>

Probability that one of the resources from the functional configuration $N^{(i)}$ would fail and all K reconfiguration groups would operate correctly is estimated as

$$\begin{aligned}&\left(f^{Case2} \in \left(F\left(N^{(i)}=1\right) \cap F\left(REC=0\right)\right) \Rightarrow\\ &\Rightarrow P_{Case2}^{(i)} = N_C^{(i)}\, r_S^{N_S^{(i)}}\, r_C^{\left(N_C^{(i)}-1\right)}\left(1-r_C\right)\left(r_S\, r_C^2\right)^{K^{(i)}}\end{aligned} \tag{13}$$

while the task execution time is increased by the reconfiguration time

$$\begin{aligned}\tau_{Case2}^{(i)} &\approx N_S^{(i)}\tau_{serv} + (N_C^{(i)}-1)\,\tau_{comm} + \tau_{serv} + 2\tau_{comm} + \tau_{reN^{(i)}} \approx\\ &\approx \tau_{nom\,T}^{(i)} + \tau_{serv} + \tau_{comm} + \tau_{reN^{(i)}}^{(i)}\end{aligned} \tag{14}$$

Case 3

Functional configuration of the *i*-th task is equipped with a reconfiguration group (of cardinality *K*) which implements the principle of first-order neighbourhood ($reN^{(i)} = 1$). Probability of correct task execution which takes into account possible *k* reconfigurations, with assumed independence between failures of resources in the reconfiguration group (bold assumption!), can be estimated by the sum of probabilities of correct operation of *k* ($1 \le k \le K$) paths

$$P_{Case3} = \sum_{k=1}^{K} P_{Case3,k} \approx$$
$$\approx N_C^{(i)} r_S^{N_S} r_C^{\left(N_C^{((i))}-1\right)} \left(1-r_C\right) \left(\sum_{k=1}^{K} (K-k) \left(r_S r_C^2\right)^{(K-k)} \left(1-r_S r_C^2\right)^k \right) \tag{15}$$

and the task execution time

$$\tau_{Case\,3}^{(i)} \approx \tau_{nom\,T}^{(i)} + \sum_{k=1}^{K} P_{Case3,k}^{(i)} \left(\tau_{serv} + (2k-1)\tau_{comm} + k\tau_{reN^{(i)}}^{(i)} \right) \tag{16}$$

Case 4

Because there has been at least one resource failure in the functional configuration of the *i*-th task and all elements of the reconfiguration group are faulty, execution of this task will be suspended for the time of network renewal i.e. for $\tau_{REN}^{(i)}$. In real-world conditions renewal processes for particular resources in the functional configuration and in the reconfiguration group depend not only on the number of "service technicians" and on adopted renewal strategies but also on the relation between time to failure and repair times of individual resources. Some assessment of the lower limit of the network renewal time can be formulated after adopting minimal values of resource repair times;

$$\tau_{REN}^{(i)} \ge \min \left[Tr\left(N^{(i)}=1\right), Tr\left(REC=1\right), \cdots, Tr\left(REC=K\right) \right] \tag{17}$$

where $Tr(\ldots)$ denotes renewal time of a failed resource.

Equation (16) is valid with the assumptions that every renewal starts at the time of failure detection, there is a sufficient number of independent "service technicians" and the cumulative renewal time is much shorter than the average time between failures (so called *fast recovery*). The last condition is entirely reasonable in the case of contemporary computer equipment which usually suffers from single failures over the period of two or three years of continuous operation [5, 9, 14].

4.3.3 Number of Network Renewals

Over the analysed exploitation period *T* the network carries out tasks with cumulative execution time $T^{(i)}$. By assumption, the time devoted to realization of the reconfiguration procedures is included in the task execution time;

$$\mathrm{T} = T^{(i)} + L_T^{(i)} E \tau_{REN}^{(i)} \tag{18}$$

where

$L_T^{(i)}$ - number of network renewals which result from inability of reconfiguration,

$E\tau_{REN}^{(i)}$ - expected value of renewal time of i-th functional configuration.

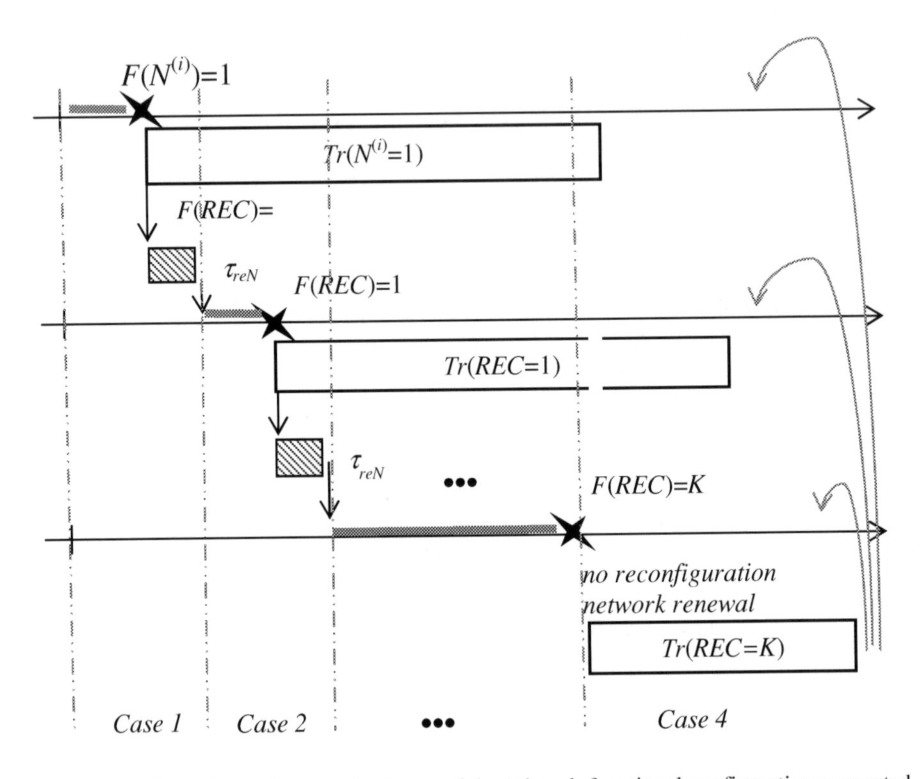

Fig. 4. Reconfiguration and renewal schema of the i-th task functional configuration supported by the reconfiguration group

With an assumption that the i-th functional configuration engage $N^{(i)}$ network resources and that the reconfiguration group implements the principle of first-order neighbourhood, the network renewal takes place with every transition of the system into *Case 4*, i.e.

$$T = \mathrm{T}_T^{(i)} + T_{lost}^{(i)} = \sum_{m=1}^{3} P_{Case\ m}^{(i)}\, T\, \tau_{Case\ m}^{(i)} + L_{lost\,4T}^{(i)}\, E\tau_{REN}^{(i)} \tag{19}$$

or also

$$T = \mathrm{T}_T^{(i)} + T_{lost}^{(i)} = \sum_{m=1}^{3} L_{Case\ mT}^{(i)}\, \tau_{Case\ m}^{(i)} + L_{Case\,4\,T}^{(i)}\, E\tau_{REN}^{(i)} \tag{20}$$

where

$L^{(i)}_{Case\ mT}$ - number of state transitions into *Case m* of the network operation within the analysed exploitation period T.

If the *i*-th functional configuration is considered as a system with as alternative renewal process [6, 5] then the expected number of renewals (understood as returns from *Case 4* – see Figure 4) can be estimated as

$$L^{REN\,(i)}_{Case\,4\,T} \approx \frac{\Lambda^{(i)}_{Case\,4} M^{(i)}_{REN}}{\left(\Lambda^{(i)}_{Case\,4} + M^{(i)}_{REN}\right)^2} * \\ * \left[-1 + \left(\Lambda^{(i)}_{Case\,4} + M^{(i)}_{REN}\right)T + \exp\left(-\left(\Lambda^{(i)}_{Case\,4} + M^{(i)}_{REN}\right)T\right)\right] \quad (21)$$

whereas the expected number of network failures understood as the sum of state transitions into *Case 4* as

$$L_T^{f(Case\,4^{(i)})} \approx \frac{\Lambda^{(i)}_{Case\,4}}{\left(\Lambda^{(i)}_{Case\,4} + M^{(i)}_{REN}\right)^2} * \\ * \left[\Lambda^{(i)}_{Case\,4}\left(1 - \exp\left(-\left(\Lambda^{(i)}_{Case\,4} + M^{(i)}_{REN}\right)T\right)\right) + M^{(i)}_{REN}\left(\Lambda^{(i)}_{Case\,4} + M^{(i)}_{REN}\right)T\right] \quad (22)$$

If we assume a long exploitation period $T >> MTBF_{Case\ 4} >> L_T^{REN^{(i)}} \tau^{(i)}_{REN}$ then the number of renewals is estimated as

$$L^{REN\,(i)}_{case\,4T} \approx \frac{\Lambda^{(i)}_{Case\,4} M^{(i)}_{REN}}{\left(\Lambda^{(i)}_{Case\,4} + M^{(i)}_{REN}\right)^2}\left[-1 + \left(\Lambda^{(i)}_{Case\,4} + M^{(i)}_{REN}\right)T\right] \quad (23)$$

or even further approximation is allowable $L^{REN\,(i)}_{Case\,4T} \approx \frac{\Lambda^{(i)}_{Case\,4} M^{(i)}_{REN}}{\left(\Lambda^{(i)}_{Case\,4} + M^{(i)}_{REN}\right)} T$, whereby the difference between the expected numbers of failures and renewals over the period $[0, T]$ is

$$L_T^{f(Case4^{(i)})} - L^{REN(i)}_{Case4T} \approx \frac{\Lambda^{(i)}_{Case4}}{\Lambda^{(i)}_{Case4} + M^{(i)}_{REN}}\left(1 - \exp\left(-\left(\Lambda^{(i)}_{Case4} + M^{(i)}_{REN}\right)T\right)\right) \quad (24)$$

and it asymptotically approaches the value of $\frac{\Lambda^{(i)}_{Case4}}{\Lambda^{(i)}_{Case4} + M^{(i)}_{REN}}$.

4.3.4 Estimation of Renewal Parameters

According to the statements made about uniformity of the resources and availability of the "service technicians" it is assumed that $M^{(i)}_{REN} \approx 1/\tau^{(i)}_{REN}$ wherein $\tau^{(i)}_{REN}$ can be estimated as the minimal value of resource renewal time (17).

Finding intensity of state transitions into *Case 4* is an unrealistic task in case of generically formulated prepositions about the network because it constitutes a renewable system with a moving reserve and, as it is known, transition intensities (failures, renewals) in such systems are functions of time. For the needs of an engineering approximation one can assume

$$\left(1+K^{(i)}\right)\left(\lambda_S+2\lambda_C\right)\leq \Lambda^{(i)}_{Case\,4}\leq\left(N^{(i)}+K^{(i)}\right)\left(\lambda_S+2\lambda_C\right) \tag{25}$$

where the left-hand limit corresponds to a situation when there is one failure of an element in the basic functional configuration and the failures of all elements in the reconfiguration group, and the right-hand one allows a failure of any one element in the functional configuration and in the reconfiguration group (reliability serial structure).

$$\frac{\left(1+K^{(i)}\right)\left(\lambda_S+2\lambda_C\right)}{\tau^{(i)}_{REN}\left(1+K^{(i)}\right)+1}T\leq L^{REN^{(i)}}_{Case4T}\approx L^{f(Case4)}_{T}\leq\frac{\left(N^{(i)}+K^{(i)}\right)\left(\lambda_S+2\lambda_C\right)}{\tau^{(i)}_{REN}\left(N^{(i)}+K^{(i)}\right)+1}T \tag{26}$$

For simplicity it was assumed that the average intensity of state transitions into *Case 4* corresponds to

$$\Lambda^{(i)}_{Case\,4}\approx\left(\lambda_S+2\lambda_C\right)\left(\frac{N^{(i)}+1}{2}+K^{(i)}\right) \tag{27}$$

and

$$L^{REN^{(i)}}_{Case\,4T}=L^{f(Case\,4)}_{T}\approx\frac{\left(\frac{N^{(i)}+1}{2}+K^{(i)}\right)\left(\lambda_S+2\lambda_C\right)}{\tau^{(i)}_{REN}\left(\lambda_S+2\lambda_C\right)\left(\frac{N^{(i)}+1}{2}+K^{(i)}\right)+1}T \tag{28}$$

4.3.5 Reconfiguration Effectiveness Factor

For the nearest neighbourhood principle the reconfiguration effectiveness factor defined in (4) becomes

$$\eta_T(reN^{(i)}=1)\approx\frac{\tau^{(i)}_{nom\,T}}{\tau^{(i)}_{realT}(reN=1)}\left(1-\frac{T^{(i)}_{lost}(reN^{(i)}=1)}{T}\right) \tag{29}$$

where

$\tau^{(i)}_{nom\,T}$ - the average nominal execution time for the i-th task evaluated as (12),

$T^{(i)}_{lost}\left(reN^{(i)}=1\right)\approx L^{REN^{(i)}}_{Case\,4T}\tau^{(i)}_{REN}$ - cumulative lost time of the i-th functional configuration,

$\tau^{(i)}_{real\,T}\left(reN^{(i)}=1\right)=\tau^{(i)}_{real\,T}\approx\sum_{m=1}^{3}P^{(i)}_{Case\,m}\,\tau^{(i)}_{Case\,m}$ - the average actual execution time of the i-th task estimated in the analyses of *Case 1*, *Case 2*, *Case 3* and *Case 4*.

Probabilities $P^{(i)}_{Case\,m}$ and times $\tau^{(i)}_{Case\,m}$ are given by equations (11), (13), (15) and (12), (14), (16).

Value of the reconfiguration effectiveness factor can be calculated using the above proposed expected values for transition intensities between *Case* states (27) and (28) or renewal times of the i-th functional configuration (17). Also, one can use the variability intervals proposed in (25) and (26).

4.4 Estimating Reconfiguration Effectiveness Factor: An Example

In this example we will consider a network of $N = N_S + N_C$ resources. For execution of a i-th task a functional configuration $N^{(i)}$ is established which engages $N_S^{(i)}$ servers and $N_C^{(i)}$ communication links.

Task execution time is the sum of times taken by service completions in particular servers and by communications between the servers (1). From another point of analysis, task execution time comes as an expenditure incurred by data processing and communication activities in some defined functional configuration. It is assumed that these times are small (they are comparable to service completion times in distributed computer networks) and that transmission times (τ_C) are larger than processing times (τ_S) like, for example, 0.004 vs. 0.001.

It is furthermore assumed that reliability characteristics of the network technical infrastructure are expressed by fault intensities λ_S and λ_C, and in the considered example fault intensity of the communication links is larger than the same intensity of the servers: 0.005 vs. 0.001, respectively. When a network fault arises, a renewal mechanism is started and, in order to minimize losses incurred by the renewal, a reconfiguration group of cardinality $K = 3$ and implementing the first-order neighbourhood principle ($reN^{(i)} = 1$) is activated. The three introduced reserve paths can replace three faulty communication links in the defined functional configuration. Resources of the reconfiguration group are also faulty (the paths are deactivated for the time of renewal) and in situation when there is no reconfiguration path available for a broken link a complete network renewal occurs which takes $\tau^{(i)}_{REN}$ time, e.g. 2h. The cost of network reconfiguration corresponds to the time necessary for this operation and it is estimated to be a fraction of the network renewal time $\tau^{(i)}_{REC}=m\,k^{(i)}_{reN}\,\tau^{(i)}_{REN}$, where m = 1, 2, … and k_{reN} defines complexity of the reconfiguration process, e.g. 0.1.

For a long exploitation time T = 10 000 an attempt was made to estimate an impact which a reconfiguration group has on efficiency of a network which executes tasks involving $N_S^{(i)}$ services and $N_C^{(i)}$ communications. Figures 5a and 6 show a relationship between the effectiveness factor and complexity of the executed tasks for different values of m. Figures 5b and 6b present how time losses incurred by renewals and reconfigurations depend on complexity (more precisely: cardinality) of functional configuration.

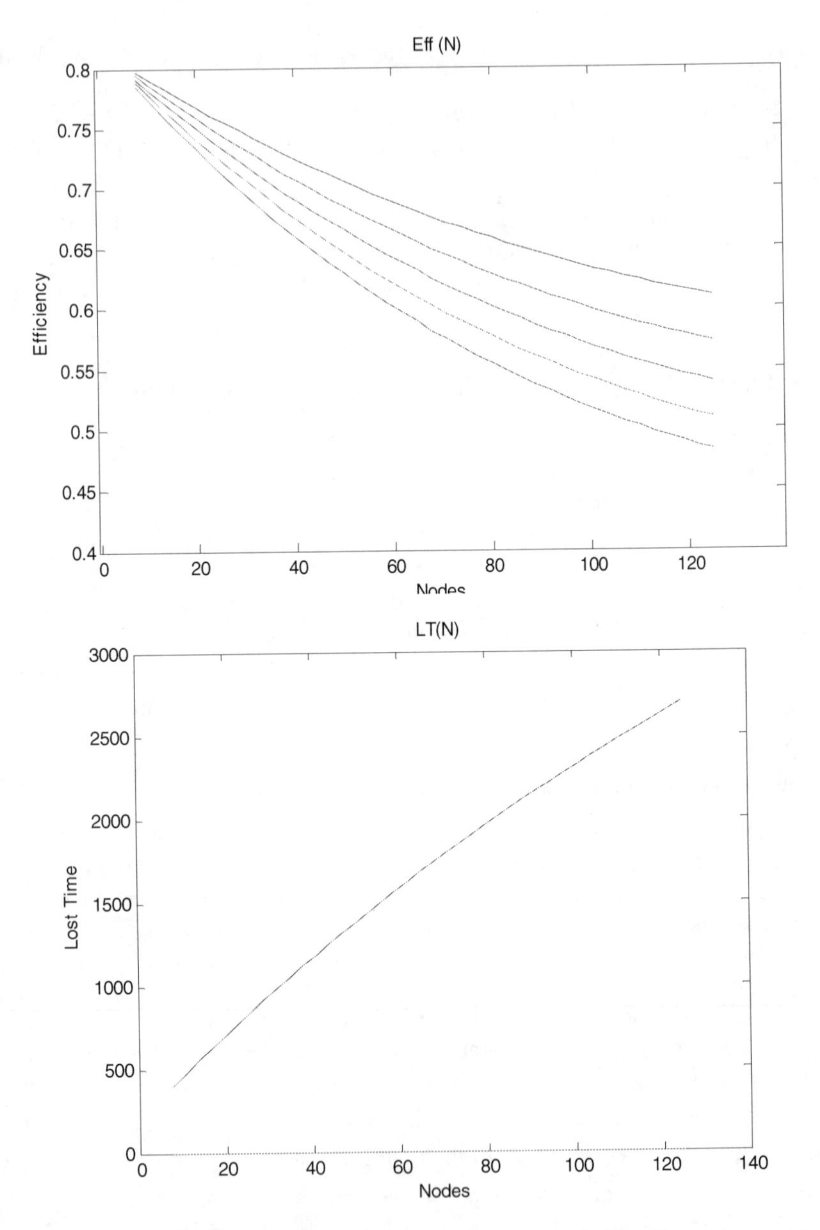

Fig. 5. Effectiveness factor (a) and lost time (b) computed for $\tau_C = 0.004$, $\lambda_C = 0.005$, $\tau_S = 0.001$, $\lambda_S = 0.001$, $\tau_{REN}^{(i)} = 2$, $k_{reN} = 0.1$, $m = 1\ldots5$

The computations were based on an average number of transitions into *Case 4* which was estimated with equation (28) and, therefore, these are initial engineering assessments only. Much more precise results could be obtained with approximations based on system renewal theory; estimates (21), (22) or even (23).

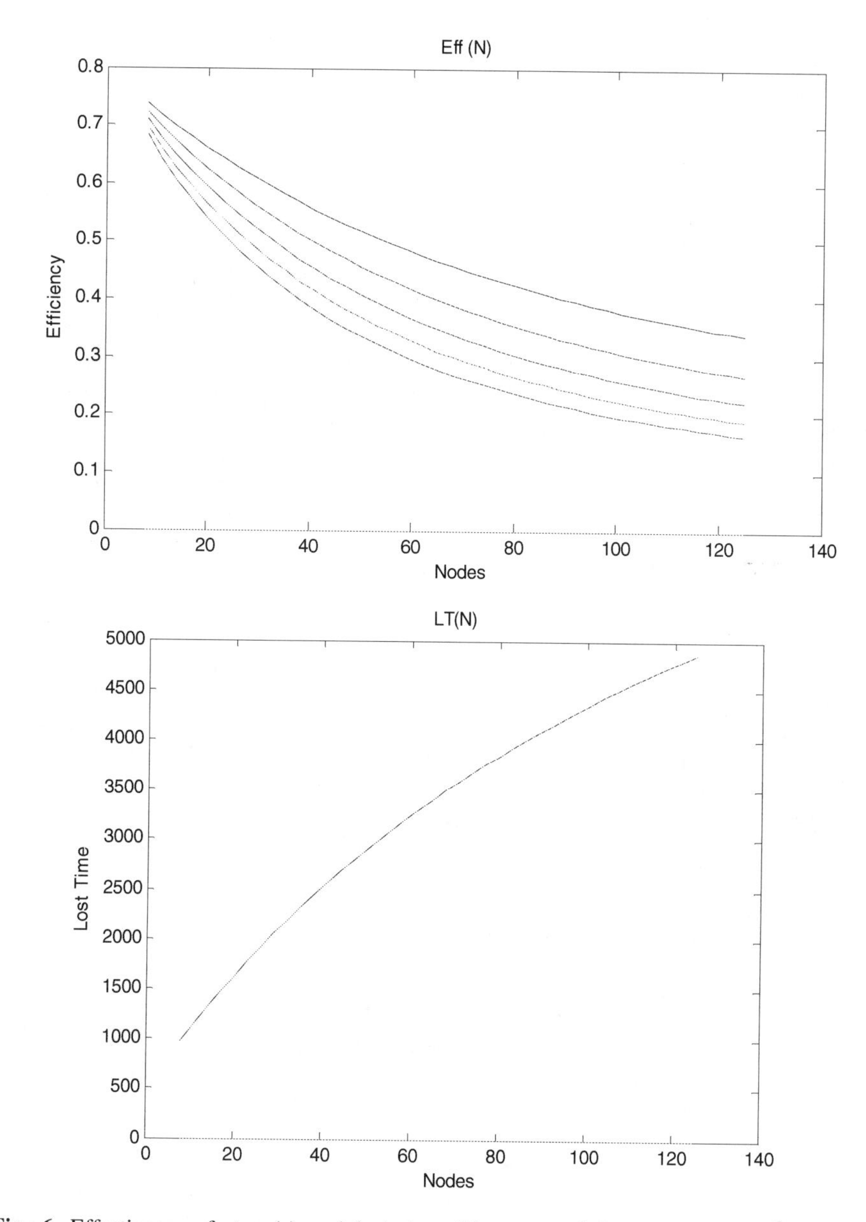

Fig. 6. Effectiveness factor (a) and lost time (b) computed for $\tau_C = 0.004$, $\lambda_C = 0.005$, $\tau_S = 0.001$, $\lambda_S = 0.001$, $\tau_{REN}^{(i)} = 5$, $k_{reN} = 0.2$, $m = 1 \ldots 5$

5 Conclusions

The functional-reliability model for services networks with reconfiguration of communication links presented in this work is a first step only and needs further development.

Possible extensions should be first of all aimed at limitations brought by assumptions about sequential nature of service execution. Also, parallel realization of transport services in a communication link used by more than one task should be taken into account. Another direction of the research would be how the limited throughput of the communication links leads to deadlock conditions (identification of communication bottle-necks).

The proposed network reconfiguration model calls for extensions which would incorporate higher orders of node neighbourhood, e.g. $reN^{(i)}$=2, 3, etc., and would indicate their impact on estimated dependability of the services network.

Particular attention needs to be devoted to computational methods which would enable at least an engineering evaluation of selected dependability parameters of a network represented according to the proposed methodology of functional-reliability analysis.

Acknowledgements. This work was prepared as a part of the Project No. N516475940 financed by the Polish National Centre for Research and Development.

References

1. Avizienis, A., Laprie, J.-C., Randell, B.: Fundamental Concepts of Dependability. UCLA-CSD Report no. 010028 (2000)
2. Caban, D., Walkowiak, T.: Improving dependability of complex information systems by fast service relocation. In: Ali, A.-D. (ed.) W: The 5th International Conference on Information Technology, ICIT 2011, Amman, Jordan, May 11-13, pp. s.97–s.101. Al-Zaytoonah University of Jordan, Amman (2011)
3. Ang, C.-W., Tham, C.-K.: Analysis and optimization of service availability in a HA cluster with load-dependent machine availability. IEEE Transactions on Parallel and Distributed Systems 18(9), 1307–1319 (2007)
4. Gold, N., Knight, C., Mohan, A., Munro, M.: Understanding service-oriented software. IEEE Software 21, 71–77 (2004)
5. Kozlov, B.A., Ushakov, I.A.: Spravocznik po rasczetunadieznostiapparatury radioelektroniki I awtomatiki. Sovietskoje Radio, Moskwa (1975) (in Russian)
6. Ross, S.M.: Introduction to probability models. Academic Press. Inc., Orlando (1985)
7. Xiaofeng, T., Changjun, J., Yaojun, H.: Applying SOA to intelligent transportation system. In: IEEE International Conference on Services Computing 2005, vol. 2, pp. 101–104 (2005)
8. Volfson, I.E.: Reliability Criteria and the Synthesis of Communication Networks with its Accounting. J. Computer and Systems Sciences International 39(6), 951–967 (2006)
9. Zamojski, W., Caban, D.: Introduction to the dependability modelling of computer systems. In: Dependability of Computer Systems DepCoS - RELCOMEX 2006, pp. s.100–s.109 (2006)
10. Zamojski, W., Caban, D.: Maintenance policy of a network with traffic reconfiguration. In: Dependability of Computer Systems DepCoS - RELCOMEX 2007, pp. 213–220 (2007)
11. Zamojski, W.: Model funkcjonalno-niezawodnościowy systemu komputer-człowiek. w: Inżynieria komputerowa. Praca zbiorowa pod redakcją Wojciecha Zamojskiego. WKiŁ, Warszawa (2005) (in Polish)

12. Zamojski, W.: Dependability of services networks. In: 3d Summer Safety and Reliability Seminars, SSARS 2009, Gdańsk-Sopot, July 19-25, pp. 387–396 (2009)
13. Zamojski, W., Walkowiak, T.: Services net modeling for dependability analysis, pp. 1–17. In-Tech (2010)
14. Zamojski, W., Mazurkiewicz, J.: From reliability to system dependability – theory and models. In: Kołowrocki, K., Soszyńska-Budny, J. (eds.) Summer Safety and Reliability Seminars, SSARS 2011, Gdańsk-Sopot, Poland, July 03-09, vol. 1, Polish Safety and Reliability Association, Gdynia (2011)
15. Zheng, Z., Lyu, M.R.: Collaborative Reliability Prediction of Service-Oriented Systems. In: 3rd Information Survivability Workshop (ISW 2000), Boston, Massachusetts, USA (2000)
16. Zhu, J., Zhang, L.Z.: A Sandwich Model for Business Integration in BOA (Business Oriented Architecture). In: Proceedings of the 2006 IEEE Asia-Pacific Conference on Services Computing, APSCC, pp. 305–310. IEEE Computer Society, Washington, DC (2006)

Author Index